WHEN THE BOSQUE RAN CLEAR

WHEN THE BOSQUE RAN CLEAR

Life Along the River from Prehistory to the Civil War

Dan Young

JILL ALFORD, RAINEY MILLER, AND TAD LAMB

Texas A&M University Press | College Station

Copyright © 2025 by Dan Young
All rights reserved
First edition

∞ This paper meets the requirements of ANSI/NISO Z39.48-1992 (Permanence of Paper). Binding materials have been chosen for durability. Manufactured in the United States of America.

Library of Congress Cataloging-in-Publication Data

Names: Young, Dan, 1942– author.
Title: When the Bosque ran clear : Life along the river from prehistory to the Civil War / Dan Young.
Description: First edition. | College Station : Texas A&M University Press, [2025] | Includes bibliographical references and index.
Identifiers: LCCN 2024058602 (print) | LCCN 2024058603 (ebook) | ISBN 9781648432910 (hardcover) | ISBN 9781648432927 (ebook)
Subjects: LCSH: Human ecology—Texas—Bosque River Valley—History. | Stream ecology—Texas—Bosque River—History. | River settlements—Texas—Bosque River Valley—History. | Human beings—Effect of environment on—Texas—Bosque River Valley. | Landscape changes—Texas—Bosque River Valley. | Cultural landscapes—Texas—Bosque River Valley—Pictorial works. | Bosque River Valley (Tex.) —Environmental conditions.
Classification: LCC GF63 .Y68 2025 (print) | LCC GF63 (ebook) | DDC 976.4/284—dc23/eng/20250210
LC record available at https://lccn.loc.gov/2024058602
LC ebook record available at https://lccn.loc.gov/2024058603

Contents

Preface
vii

Acknowledgments
xi

Chapter 1
The Birth of a River
1

Chapter 2
The First Texans
8

Chapter 3
The Early to Middle Archaic
23

Chapter 4
The Middle Archaic
35

Chapter 5
The Wet Centuries
44

Chapter 6
Social Disruption and Change
57

Chapter 7
Climate Change
73

Chapter 8
Disease Changed Everything
84

Chapter 9
The Comanches and Euro-Americans
94

Chapter 10
Conquest and Commodity
115

Notes
139

Bibliography
185

Index
215

Preface

I taught history in public school and community college for nearly fifty years, always taking more time to discuss ancient Texas history than students or administrators thought was necessary. As a child I collected interesting rocks, and with my first arrowhead I focused on those rocks associated with camp debris of early Texas Natives, especially flint. Archaeologists, both professional and avocational, call it chert. It is hard, fine-grained cryptocrystalline stone, chemically formed in some limestones. This is the stone that is flaked into tools and points, leaving flakes to be plowed up, indicating occupation by these mysterious peoples. This was the 1950s in North Central Texas, when very little land was covered with Bermuda as people were still trying to grow things without irrigation, and nobody seemed to care about a boy poking around on their property. I was obsessed with walking the fields every chance I had, picking up the colorful flakes and occasional artifact. And not just flint, there were hammerstones, with clear battering marks from their use in the first stages of chert projects. There were always sandstone *metates* and *manos*, mostly broken by the plow, whose use as grinding tools was clear. There were burned rocks that I would later understand to be very informative about the cooking activities in those camps.

I had acquired some understanding from other enthusiasts about which fields might be productive by looking for the burned rock scatter, so evident in the bare fields. This was information that would come in handy later. There was only one identification guide in those days, so typology was a matter of handling the points in the company of the few people with lithic knowledge. *Lithics* is one of the terms I was unaware of until I began to associate as a volunteer with actual archaeologists when I became a volunteer with the Gault School investigators. That's when I started to use the term *chert* and talk about lithics. I had not collected for decades because of teaching and because of the Bermuda-grass covering on the familiar surface sites. But that was fine because I was past that stage, and it was a good thing too, since I found that you should never admit to having been a collector to actual archaeologists—it's something akin to treasure-hunting and sure to raise their eyebrows and stall the conversation.

I thought I was loaded with information when I started this book in the 1970s. But I had to throw away several false starts when I discovered that I really didn't know very much. When my association began with Gault archaeologist Dr. Steve Howard and his dig teams, I began to realize the lack of depth of my understanding of early Native history. I began to meet and learn from avocational archaeologists like Del Bartlett, Steve Davis, and Phil Afford. Geoarchaeologist Dr. Charles Frederick loaned me books, taught me how to knap and how to read the soil, and doubled my knowledge about archaeology. As my understanding of archaeology and ancient Texas history grew, I began to realize how little is known about the Bosque River's neglected role in the larger story. So, I started the book project again and put together the following ten chapters.

I avoided the complex calculations typically used in the literature to refer to time passed. Archaeological papers reckon dates based on ice cores, sedimentary layering, cave deposits, tree rings, volcanic eruptions, and the decay of radiocarbon isotopes and express the time passed as BP or calibrated BP. I have converted BP to simply refer to years before present. Beginning at 2000 years ago, I use the standard AD.

The first chapter discusses the natural history of the North Central Texas Bosque River that flows from the Western Cross Timbers through the Blackland Prairies into the Brazos River at Waco. The geological processes and climate fluctuations that shaped the river's ecosystem provided an attractive hunting and gathering landscape for Native peoples. Chapter 2 reviews the most credible explanations of how the New World was populated during the Pleistocene, the lifestyles of the earliest Texans, and the hypotheses surrounding the catastrophic close of the era known as the Younger Dryas Event. Chapter 3 surveys the Early Holocene archaeological divisions known as the Early and Middle Archaic periods. The climate was warming and drying, so game, especially bison, was more difficult to find. Indigenous peoples adapting to nutritional stress developed hot-rock baking to process otherwise toxic bulbs and turned to a diet of low-carb, high-fiber desert foods, which altered their metabolism and health.

Chapter 4 further illustrates the importance of climate change in human history. During the Middle Archaic, 6,000–4,000 years ago, there was a rapid cooling event that attracted bison back into Texas southward to the Coast. The bison brought with them the Calf Creek hunting lifestyle with unique stone technology, but their prominence faded with the next shift in climate. Aboriginal peoples that frequented the Bosque River valley returned to the small group foraging lifestyle that relied on plant foods, including mesquite beans and acorns. Chapter 5 further examines the foraging life in the absence of bison along the Bosque River. These were the wet centuries after 5,000 years ago, with a general increase in population. Some of the important foods, like acorns, pecans, honey locust, cattail, and the most important camas bulbs were baked in the rock ovens. Then the climate changed again and the bison returned, along with Plains influences and a hunting lifestyle as shown in projectile point changes.

The climate has driven the bison north again by chapter 6 as catastrophic changes known in Europe as the Dark Ages bring demographic collapse as well as changes in lifestyle and territorial arrangements among Central Texas Natives. In this chapter I discuss the arrival of the bow technology, dominance of the Scallorn arrowpoint, and the impact of the Medieval Warm Period. Chapter 7 continues the discussion of the effects of the Medieval Warm Period and the historical climate changes that transformed the wet meadows along the Bosque River by droughts and floods. A cooling climate causes reintroduction of bison, which gives rise to the Toyah culture and begins the long journey of the Wichita people to the Bosque River.

History accelerates in chapter 8 as the arrival of the Spanish marks the beginning of the end for the Wichita people as European diseases and conflicts drive them to Texas. Even worse for the Native peoples were the arrivals of the Anglo-Americans and the even more violent Scots Irish, whose sense of divine purpose excused their xenophobic culture's drive toward ethnic cleansing and extermination. Treaties were ignored and there were clashes involving Cherokees, Wichitas, and Tawakonis.

Chapter 9 describes the Comanche origins brought about by climate-induced migrations and the acquisition of horses, allowing them to become a dominant predatory society and lords of the Southern Plains. A series of volcanic eruptions and resulting climatic effects led to the westward migration of well-armed Anglo-Americans into Texas on a collision course with the Comanches. Incidents such as the Parker's Fort raid, Council House, and Plum Creek, combined with constant threat from disease, mark the Comanche decline.

Beginning with the short-lived community on the Bosque River, chapter 10 chronicles the last days of Texas Natives, after a brief attempt to maintain a Texas reservation, with conflicts involving Rangers and other adventurers eventually causing them to be placed on Oklahoma reservations. I use the term *Anglo* to describe the settlers from the Deep South, brought in by Stephen F. Austin and Scots Irish leaders to characterize the illegal immigrants from the Upper South. Eventually I began to refer to all Euro-American settlers as Anglos as the two began to merge. These groups followed the narrative that justified ethnic cleansing as religious-sanctioned removal of the "depredating savage," illustrated by the establishment of Stephenville on the Bosque River as a Scots Irish settlement, which in only four years became the center of anarchy and violence on the Texas frontier. By the time the Civil War began in 1861, the pattern of vigilante violence had become common in the region for decades. Once in Oklahoma, many of the Comanches and other Natives turned to crisis cults to transition to reservation life as farmers and ranchers. Along the Bosque River, the frontier lifestyle faded and was replaced with destructive farming and ranching practices that destroyed both the communities and the land.

Acknowledgments

I would first like to thank Dr. O. A. Grant, first my college professor, then mentor, then friend. I was among those students he inspired to look closely into history and read everything I could get my hands on. The world is a better place because of him. After decades of false starts the project was renewed when I began to participate in archaeological digs in the company of Lee and Mike Flowers. Thanks, guys. And thanks to Dr. Charles Frederick, a flowing fountain of archaeological and geological information whose encouragement helped place this book back on track. I would like to thank my thoughtful daughters, Sarah and Megan, whose proofreading and patience in listening to my stories and trivia inspired me to stay with the task. And thanks to my wife, Holly Lamb, whose patience and many hours at the computer made this book possible. Thank you Alan Nelson and Dahna Branyan whose research, editing, and proofing made this book far more readable than it might have been. And thanks to Rainey Miller whose long hours of work produced most of the botanical drawings in the book. And thank you Jill Alford for your animal and plant illustrations. Thanks to guest illustrator Tad Lamb for the fish hook drawing.

WHEN THE BOSQUE RAN CLEAR

1

The Birth of a River

THE WESTERN CROSS TIMBERS

The Bosque River rises in the Western Cross Timbers of North Central Texas and flows through the Southern Great Plains prairie, underlain by limestone and running into the Northern Blackland Prairie at the confluence with the Brazos River in Waco. The Cross Timbers consist of sandy bands of oak archipelagos intruding through black clay soil-based prairie. Until sixty-five million years ago, North Central Texas was on the southern edge of the North American Inland Sea. The ebb and flow of this Cretaceous sea left sediments that became layers of limestone rock and sandstone. Limestone, along with rich organic matter, eroded into the black clay soil favored by grassland. The sandstones eroded to sandy, neutral-to-acidic soils that drained fast enough to be inviting to oaks.[1]

During the Cretaceous, the ancient shore of the Gulf of Mexico was only a hundred miles south of Stephenville. Parallel drainages then flowed more eastward than to the south. About thirty million years of erosion nearly obliterated the old limestone (except for mesas like the Chalk Mountains) before the drainage finally shifted, creating the Bosque River channel. The surface limestone rocks found along the Bosque River are relics of this water activity as well. There are fragments of fossilized wood, mostly conifers, also left over from 100 million years ago.[2] By twenty-three million years BP, life forms, including the Western Cross Timbers oaks, had evolved into their modern species. The change in mammals was even more profound, as they filled the void left after the dinosaur extinction sixty-five million years ago with a diversity of near-modern species.[3]

The last two million years, known as the Pleistocene, was a time of cooler glacial periods (stadials) and warmer interglacials (interstadials).

Fossil pollen studies identify the large trees that filled the Bosque drainage as beech, maple, hickory, pine, spruce, and several oak species generally found in the less drought-prone woodlands northeast of Texas today. One woodland component, spruce, indicates that the full glacial periods were about 10°F cooler than interglacial temperatures.[4] The presence or absence of animals known to prefer specific types of vegetation shows that North Central Texas was a pulsating mosaic of forest and prairie over these millennia. The full glacial, closed-canopied woodlands dominated the pollen record; then, with the warmer interglacial, grasses increased again.[5]

During the last glacial period (14,000 to 10,000 BP) the Central Texas prairies between the bands of Cross Timbers transitioned into an oak savanna and remained free of forest cover. The late glacial period continued to warm and dry, removing boreal conifers and allowing the more rugged oaks to dominate the Cross Timbers.[6] As the glaciers disappeared from the northern part of the continent and warming continued, fire became a major variable in shaping North Central Texas. The mesic conditions of moderate precipitation gave way to a warmer-drier environment, leaving drought-stressed vegetation that depended on earlier Pleistocene conditions to burn away.[7] The bare landscape was then filled with drought-tolerant newcomers who took advantage of the vacancies to migrate northward.[8]

The Resilient Post Oak

There were twenty-five climate oscillations between stadials and interstadials during the Late Pleistocene caused by wobbles in the earth's tilt, volcanism, and/or a massive release of methane. The interstadials were sudden; in less than a decade the mid-latitudes warmed 12.6°F for a few millennia followed by gradual cooling back to wetter times.[9] The warmer drier times encouraged the post oaks (*Quercus stellata*) to temporarily dominate over more temperate trees, likely reducing those to a residual population. Later the return to a cool moist climate allowed other temperate broadleaf hardwoods to once again dominate the landscape. This assumes they survived the millennial-length interstadial and again crowded the oaks, pushing them back into minority status.[10]

The abrupt, warm interstadials resulted in times of trial, when the genetic line narrowed, typically the end of the line for trapped plants, unless they were able to adapt through a hardy subspecies with some advantage

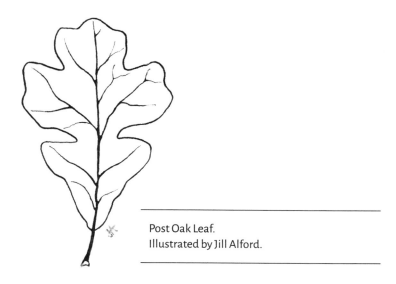

Post Oak Leaf.
Illustrated by Jill Alford.

lacked by the other plants under stress. For post oaks, this was smaller in size, with only a few large limbs, and more narrow leaf lobes, an adaption frequently occurring generally as flora tries to adjust to aridity. As the post oak survived each warm-up, new, better-suited subspecies populated refuges like the Cross Timbers. Additional trials refined the post oak's survivability, and it became the most fire and drought resistant of the oaks. Post oak acorn sprouting and survival occur in a break during these droughty times when the understory is sparse. For example, the generation of post oaks now reaching middle age on the upper Bosque River dates to the return of rainfall after the Civil War drought of 1855–64.[11]

There were always plenty of species dropping into the Cross Timbers poised to take hold, known as "seed rain." A variety of animals placed the seeds on a germination fast track through their digestive systems, especially robins and cedar waxwings with the oddly appropriate scientific names *Turdus migratorius* and *Bombycilla cedrorum*. These seedlings usually came to a bad end when they sprouted in thick understory growth. The post oak has been successful because it reproduces through a shock doctrine strategy of taking advantage of ecological catastrophes to advance a new generation. An observer during the long mesic periods of the Pleistocene would not have predicted that the post oak, pressed to a remnant at times, would be around much longer. But then drought, followed by lightning-started fires, would kill off the competition, and, as tree rings have shown, three or four wet years would bring up a cohort of oaks large enough to join the forest canopy. This post oak coup is called a recruitment pulse, when several oaks reach about five feet, standing free from competition. The post oak's centuries-long patience, chemical suppression of competing sprouts, and superior drought and fire resistance have proven the dictum about old age and treachery.[12]

The Mast Year Strategy

The recruitment pulse naturally follows a mast year. Mast comes from the old English term for "forest food," with mast years referring to the occasional heavy crop of acorns or pecans. In mast years the squirrels, blue jays, and mice carry off the bounty to store over the winter. Troubled with sprouting acorns in storage led to an evolved squirrel behavior of biting off the acorn's tip to kill the seed's embryo, so that the acorn lies throughout the winter in an edible state, but unable to sprout, a little like the paralyzed spider, stung and dragged into the wasp's den. The oak's response was to produce large numbers of acorns, some with offset embryos, to survive the fate of the zombie acorns created by squirrels. The masting strategy is called "predator satiation," when the rodents eat or destroy all they can, but can't help leaving plenty of nuts scattered over a wide area. If the oak produced only a few acorns each year, they would all be eaten. If every year were a dependable crop, woodland mice, squirrels, and especially insects would synchronize their lives around the crop and destroy it completely every year. Unpredictable masting avoids predator population build-up by starving the predators for the next few years. Foresters sometimes claim that they can predict a mast year, but the mechanism remains elusive, probably something to do with an environmental cue that is taken by millions of oak trees at the same time, across several species. Whatever the timing, they appear synchronized within a species with tree masts occurring simultaneously over hundreds of square miles.[13]

The Mycelium Internet

Synchronization among plant life is a troubling concept to the Western mind. The idea that something is making decisions about collective forest masting and responses to challenges in the environment does not seem rational. Yet there is an internet under the soil, the fungi mycelium cloud, a primeval arrangement in which plants pay a tax in return for double the water and nutrient supply. A case can be made for an interested, even caring force in the fungi nether-cloud. The benefits go well beyond increased nourishment; like the Force in Star Wars, the mycelium occupies the forest soil as an unseen network that transmits information as well as water and nutrition among the trees. Competing trees on the surface struggle for place, but under the soil, the ancient mycelium Force considers the whole forest ecology, not just post oaks, all members of the realm. An insect attack—the exact species recorded by its saliva—or serious browsing activity causes a disturbance in the Force that alerts the whole membership to ramp up the tannins or other toxins in the leaves. The mycelium internet transmits information by electric impulses over hundreds of yards.[14]

A wounded or sick tree, if found worthy, will become the beneficiary of nutrients sent from other forest trees—even trees that could benefit from the vacancy if that tree died. The fungus survives only as long as there is a forest, which partly explains why the mycelium works to benefit the entire forest, not just the post oaks. The "redistribution system" keeps the whole forest healthy as protection from a single catastrophic pandemic event. There is, however, a caring component that doesn't seem to have anything to do with the collective forest good. The spot where an ancient tree fell, centuries ago, sometimes has mossy knobs of the ancestral tree's stump continuing to live. The wood web has continued to pump sugars and moisture into these ancient fragments, which, without photosynthesis, could not have lived so long. This is where the tree observers start to wonder about why the "decision" was made to keep a nonproductive member of the community alive. It's almost like ancestor worship. There must be value in a ghost tree's memory.[15]

A fungal regime can be overthrown by tree-killing fungus, spreading through the forest and seeking edible tissue. Such a catastrophe is less likely on ridges where the soil is thin, between the more intense fungal wars in the rich soils and the poverty of nutrition in the thin soils. This is where the time-battered post oaks reach very old age—as long as 500 years. The oldest trees possess curious characteristics, such as missing tops, killed by lightning or disease, known as "stag tops," a spiraling of the bark up and down the trunk that even shows up in the wood grain, several old scars where branches have self-pruned, breaking off cleanly, because the limbs are so brittle. There are only a few limbs on the most ancient post oaks, but those are large, and gnarled, forming a very open canopy. They are not the largest of post oaks, but they are always the oldest trees in the vicinity.[16]

There are subordinate species that follow the post oak's lead to establish a balanced community. During xeric periods, post oak's shorter-lived, rugged cousin, the blackjack oak (*Quercus marilandica*), can outlast the competition by holding out on the least fertile soil. Another blackjack adaption is its ability to sprout after a fire from a top-killed sapling. But when the climate returns to mesic conditions, the blackjack lacks resistance to crowding and is among the first to be eliminated by moisture-loving hostile fungal and insect activity.[17] The cooler, wetter spells favor the large bur oak (*Quercus macrocarpa*). The crown of this tree can be over a hundred feet tall. Tolerating both moisture and a little clay in the soil, this tall tree occupied the floodplains, dropping two-inch acorns in the fall.[18]

The pecan (*Carya illinoinensis*) was a later arrival that waited out most of the Pleistocene in Mexico before breaking with the other hickories, becoming cold-adapted and migrating into the Cross Timbers bottomlands. Along the way, the pecan's evolutionary companion, the fox squirrel (*Sciurus niger*), accelerated the spread northward. The most interested avian dispersal agent was the blue jay (*Cyanocitta cristata*) and a few others, none of which were advancing the development of larger nuts with thinner shells. It was the crow (*Corvus brachyrhynchos*) that was large enough and selective enough to prefer larger, thin-shelled nuts, before flying miles for an accidental drop. Pollen records trace this migration over the last 20,000 years,[19] during which human preference for shell thinness and taste was likely a factor as the pecan became an important food.[20]

At the close of the Pleistocene, as the climate was warming, an important food tree migrated out of the Mexican deserts and into the Western Cross Timbers, the honey mesquite (*Neltuma glandulosa*). The mesquite had originated as a tropical tree that later found itself trapped in the Sonoran Desert for millions of years where aridity-driven evolution produced the Texas tree.[21] During the mesquite's xeric isolation, it developed long—up to ten inches—sugary pods to tempt now-extinct mammals to act as dispersal agents. Elephant-like gomphotheres, ground sloths, camels, and horses benefited from the mesquite pods but were deterred from eating the foliage by thorns and toxins developed to protect the leaves and twigs.[22] When modern horses and cattle eat and spread the mesquite seeds, they are continuing the Pleistocene dispersal strategy.

Unlike the other components of the Western Cross Timbers, the mesquite is even more dependent on fire. An adult tree is rarely killed by burning, and the reproductive strategy depends on grass fires or drought to clear the soil for sprouting, sometimes miles away from the tree as the seeds pass through an animal's digestive tract.[23] Sometime around the end of the Pleistocene, the honey mesquite became a minor part of the Western Cross Timbers.[24] Mostly these trees, oblivious to soil type, grew as the most visible component of a mesquite savanna that was maintained by grassfires that regulated the mesquite's spacing.[25]

The Western Cross Timbers includes fire-adapted companion plants long associated with post oaks that occupy the forest understory. There are brambles, which include the notorious, spiny greenbrier (*Smilax bonanox*) and the thornless tangles of snailseed vine (*Cocculus carolinus*) with its showy clusters of red berries that prefer the same sandy soil required by post oaks. Greenbriers are probably more common and chaotic today than when once numerous beavers delighted in consuming the tubers near streams, restricting the briar—and most trees—several yards away from water. The prickly dewberry canes (*Rubrus trivialis*),[26] the hanging Virginia creeper (*Parthenocissus quinquefolia*), and the spiny chittamwood bush (*Bumelia lanuginosa*)[27] add to the impenetrable understory chaos so often described by early travelers.

Early Spanish *entradas* used the terms *monte grande*, which can mean "large forest," and *monte del diablo*, in reference to the vine-twisted undergrowth. In places the forest required travelers to get off of their horses and chop their way through the dense undergrowth of the "Cast Iron Forest,"[28] yet in numerous other locations, the timber was open enough to drive wagons through in any direction.[29] The large parklike areas were most likely maintained by purposeful burning carried out by American Natives to attract game, clear the ground for collecting acorns and pecans, destroy ticks, and protect the blackberries from disease.[30]

The Late Pleistocene Western Cross Timbers and prairies provided a tempting biome that was home to a fantastic array of mammals. The First Peoples to enter this area were probably following migratory animals like the Columbian mammoth (*Mammuthus columbi*), drawn to the rich grasses, the browsing American mastodon (*Mammut americanum*), and several species of horses (*Equus semiplicatus*), giant bison (*Bison antiquus*), and camels (*Camelops macrocephalus*).[31] The relationship between people and animals that would characterize later life along the Bosque River was soon established; man and beast were migratory, one followed the other.

2

The First Texans

People had been migrating out of Africa for more than 100,000 years, spreading over the globe before they settled in Beringia around 30,000 years ago;[1] they had no idea that they occupied the gateway to the Americas. Beringia did not look like a gateway or a land bridge; it was a Texas-sized land mass six hundred miles wide covered with shrubby tundra and loaded with elk and smaller game. It was a lost continent. The glacial gateway that plugged the American end of Beringia opened several times over millions of years on a millennial scale, because of climate oscillations. It allowed for the shuttling of plants and animals back and forth across the occasional ice-free corridor.[2]

The Beringian Corridor

The standard explanation of how people moved from Siberia to North America, the one found in most textbooks, is that at the end of the Pleistocene, when the ice melted enough for people to pass through an icy corridor to the New World around 13,000 years ago, migration occurred. More recent investigations cast doubt on that theory because there are so many North American archaeological sites with firm, older dates.[3] The problem is that migration was held up for millennia when the final, most intense surge of Pleistocene cold drove Beringians out of known archaeological sites. This was the Last Glacial Maximum (LGM), which peaked about 20,000 years ago. Paleoclimatologists have studied sediment cores from this period and found pollen, insect remains, and plant fossils that show some of the LGM environments to have been dry and cold grasslands, supporting horses, bison, and mammoths. The missing East Asian Siberians might have been able to survive somewhere to the south, along the coast. Though the area is now covered with 164 feet of water, the southern-central part of Beringia could have been warmed by the ocean, with an attractive environment.[4]

Evidence is hard to find under that much water, yet several genetic studies show that wherever they were, populations became isolated during the harsh LGM. The group that became the ancestral First Americans lived in a refugium for 5,000 to 8,000 years, long enough for their DNA to drift away from their Asian parent groups. This period is called the "Beringian Standstill," a place where they waited for the LGM to thaw enough to open the corridor between the Laurentide and Cordilleran glaciers, the gateway to North America.[5] Except that's not what happened; recent research shows that such a corridor would have been impassable for thousands of years because there would have been hundreds of miles without vegetation or animals, and the passage would have contained ice boulders, lakes, and mud pits.[6] It's just not likely that people could have followed melt rivers as they cut an opening in the ice between the glaciers.[7] There is no archaeology marking the arrival of these people through an ice corridor, though there are those intriguing Native stories about the First People arriving on earth through a hole in the ground—(I have wanted to believe this was a reference to a river-cut ice tunnel). The timing is just too late.[8]

The Kelp Highway

For decades the accepted date for the opening of the ice corridor was between 13,000 and 15,000 years ago. Recent scholarship throughout the Americas has shown that there were well-established pre-Clovis populations 16,000 or more years ago. The Kelp Highway, the Pacific Coastal route, is now accepted by many geneticists and archaeologists as likely the first route to be ice-free. Geneticists insist that there had to be a pause somewhere, and this proposed route doesn't challenge that the "standstill" took place in southern Beringia during the fierce Last Glacial Maximum. If this millennia-long "incubation" was a coastal environment, it was a good place to develop a taste for kelp and acquire the skills to hunt and fish in a coastal biome as well as the nearby mammoth steppe. This route southward, opened 17,000 years ago or even earlier, would have been faster than moving south through the American interior, where constant ecosystem changes with each latitude

Columbian mammoth, *Mammuthus columbi*. Mammoths evolved in Africa about five million years ago, and arrived in North America about 1.5 million years BP. The Columbian mammoth became the most widespread form on this continent, standing 13 feet at shoulder height. The woolly mammoth, *Mammuthus primigenius*, developed in Beringia/Siberia around 100,000 BP and spread into North America and Europe. Much of what is known about the Columbian mammoth is derived from the bones and teeth from a collection of Pleistocene mammals, including twenty-three Columbian mammoths at the confluence of the Bosque and Brazos Rivers who died there around 67,000 years ago. During a glacial interval (71,000–57,000 BP) that was hotter and drier than today, dozens of animals converged at the last area waterhole near Waco (Waco Mammoth National Monument), where dentition reveals that they drank highly evaporated water for months before perishing from thirst.

Donald A. Esker, "Thermoregulation and Dental Isotopes Reveal the Behavior and Environment of Pleistocene Megafauna at Waco Mammoth National Monument," PhD dissertation, Baylor University, 2019; Bjorn Kurten, *Before the Indians* (New York: Columbia University Press, 1988); Am M. Lister, et al., "Evolution and Dispersal of Mammoths across the Northern Hemisphere." *Science* (2015). Illustrated by Jill Alford.

would have taken more time. The coastal ecosystem stayed pretty much the same until the warmer waters off Baja, California, changed the climate.[9] At any of the rivers meeting the coast, the First Peoples could have followed it inland and would have been tempted to hunt the Pleistocene herds, moving with them eventually to the Great Plains and south to Texas.[10]

Pleistocene Texas

The First Texans came during a warm, moist, late glacial warm cycle of increased forest cover with prairie openings. Evidence for these conditions is inferred from sediment cores and excavations from just outside the Western Cross Timbers in the form of environmentally sensitive fossil pollen, beetles, fauna, and the latest investigative tool, phytoliths—little stone fingerprints inside plants.[11] Phytoliths, along with teeth and bone chemistry from Texas mammoths and bison, help describe the prairie openings as C3 grasses, spoiled to the lush Pleistocene climate characterized by easy transitions from season to season without sharp temperature differences. The soil supporting the Central Texas biologically diverse ecosystem, shown by the presence or absence of burrowing animals, had to have had a thickness of around 140 cm, deep enough to have supported a robust ecosystem. But their time was coming to an end.[12]

The late glacial warming accelerated 14,700 years ago, bringing more broadleaf trees with plenty of prairie gaps to attract large grazing mammals like the Columbian mammoth, the thirteen-foot-tall, eight-ton elephant that could graze on the succulent C3 grasses. Strontium isotopes in the enamel of Central Texas mammoths shows that they migrated up to 500 km each year. Grazing partners included the smaller Jefferson's mammoth, giant *Bison antiquus*, three species of horses, and two types of llamas and camels. There were plenty of animals avoiding the open grasslands in preference for shrubs and tree foliage, including browsers like the American mastodon, white-tailed deer, the pronghorn "antelope," four-toed ground sloths, a six-hundred-pound beaver, and an assortment of tapirs. Then there were the six-hundred-pound armadillo and a two-thousand-pound glyptodont that looked like a cross between a Volkswagen bug and a massive armadillo. When the First Texans entered this ecology, they had to be ready to manage the predators that followed these animals: scimitar-cats, American lions, and some saber-tooth cats, all larger than African lions.[13]

The Gault Site, 18,000 Years Ago

If these early people left any of their projectile points or tools along the Bosque River, they will not be recognizable until the pre-Clovis points at Gault have been typed. The justly famous Gault site just north of Austin contains one of the few North American occupation zones that date back more than 16,500 thousand years. An exact notion of what a pre-Clovis tool kit looks like will soon be established. The tools and dart points represent several styles and lithic traditions.[14] These people are going to remain mysterious for some time as archaeologists work to make sense of their mostly lithic remains. The variety of the stone technologies represented at this deepest level suggests that these were localized societies with possible continent-wide connections. Their chert points were made without the big-game hunter's skill or pride, and it's likely that their nutrition came more from smaller animals and gathering the plentiful Pleistocene vegetation. There are no bones of the big predators in their camp debris,[15] and it may turn out that they were avoiding the megafauna for a more peaceful coexistence. Eventually, after thousands of years, there was a new and exciting development in human history, and it is recorded at Gault, not so far south of the Bosque River.

A popular campground during most of the deep history of Texas was on Buttermilk Creek in Central Texas, known today as the Gault site. Under about twenty feet of sediment, the story of the First People, arriving before 18,000 years ago, is recorded. The creek is spring-fed, and it runs through a small valley with soccer-ball-size cobbles of fine-grade chert that can be easily dug from the stream's edge. Buttermilk Creek is near woodlands and prairie where hunting and plant-gathering could have been done with ease. Just above the early sediment begins a record of the most celebrated in early American archaeology—the Clovis tradition. The Clovis is generally dated to have begun around 13,500 years ago. The earliest of these projectile points at the Gault site are that age and attest to a record of their existence until their untimely end around 12,750 years ago. The Clovis point was used for hunting a wide range of game but is best known for confronting the megafauna of the day, especially the mammoth and giant *Bison antiquus*.[16] Perhaps the Clovis tradition is best known for their magnificent Clovis dart points, found sparingly in collections from Central America to Canada.

The Clovis lithic technology must have developed across a wide area as regional groups met through trade and compared stone technologies and hunting techniques, eventually arriving at an exciting new lifeway that featured the popular, new Clovis design. The Clovis lithic tradition is not thought to have represented a single culture, but it was probably a must-have technology, like cell phones today, that soon took shape and held its coherence for centuries. One thing that archaeologists agree on is that the Clovis points are frequently made from exotic cherts, traded from hundreds of miles from where they originated.

Imagining Clovis

The Clovis points were so important to these Pleistocene people that it has to be assumed that they were charged with symbolic meaning. Ethnographic literature provides examples of whole vocabularies dedicated to the chert cobble's interior. There were zones of low value, such as the less dense layers. Among the pure "killing stone," colors too must have had significance because some quarries show that equally knappable cherts were ignored while another color from the same site was selected. There was probably something beyond utilitarian considerations at work.[17] The modern mind keeps forgetting that spiritual considerations obsessed aboriginal life. Excavation of power chert cobbles, like those easily removed from the Buttermilk Creek bank at Gault, could have been done by those designated with the sacred task, and the processing of the stone into trade blanks (spalls reduced to bifaces) was perhaps ritual-bound. What happened to these trade blanks raises the question of whether the Edwards chert from Central Texas was spread all over the Great Plains to Canada by intrepid young men on a sacred pilgrimage, dispersed by the Clovis people as they followed the migration of totem animals, or did a continent-wide trade network develop?

A trade network, developing along with the Clovis culture, would explain the vigor, uniformity, and brilliance of their dart points and ivory work. Consider human nature and how people feel elated about a good deal. The result of coming together at seasonal locations for a trade fair, to exchange power cherts, ivory foreshafts, shells from both oceans, obsidian, and the all-important red ocher,[18] led to the equally invigorating exchange of ideas. Throughout history cultures have been amplified by developing a collective brain, the exchange of ideas that created a network that brought together information that had been discovered in isolation. Consider

The Clovis point (13,000—12,700 BP), unless resharpened, is as long as 150 mm with several long flakes removed from the base. Unlike later dart points the Clovis was manufactured from spalls struck from a polyhedral core. The indirect flaking is so fine that an occasional "overshot" flake will travel across the blade; finding these flakes usually indicates a Clovis work area. These points are associated with mammoth kill sites all over Texas.

Ellen Sue Turner, Thomas R. Hester, and Richard L. McReynolds, *Stone Artifacts of Texas Indians* (New York: Taylor Trade Publishing, 2011); Michael R. Waters, Thomas W. Stafford, and David L. Carlson, "The Age of Clovis—13,050 to 12,750 B.P." *Science Advances* 6, no. 43 (October 21, 2020); and Christopher S. Jazwa, Geoffrey M. Smith, et al., "Reassessing the Radiocarbon Date from the Buhl Burial from South-Central Idaho and Its Relevance to the Western Stemmed Tradition-Clovis Debate in the Intermountain West," *American Antiquity* (2020). Illustrated by Jill Alford.

herb-based medicine developed over centuries by isolated populations and made available through diffusion at trade fairs. Instead of knowing only what the band's healer knew, the trader tapped into the collective brain, having access to many lifetimes of learning.[19]

Tattooing may have been an important symbolic element, as evidenced by the burial of a shaman and his young female companion just a few miles from the mouth of the Bosque River in the Horn Rock Shelter. Among the hundred items found with the man was a turtle shell containing a sharpened coyote tooth, grindstones with antler pestles, and red ocher.[20] Just who was marked by what tattooed symbols is an exercise in imagination. Nothing excavated from the earth can explain the belief systems that held this Pleistocene world together. The best piece of hard evidence is the Clovis point itself. First of all, it was made with great skill; it was fragile and frequently broken in production and was made only with beautiful, power-laden chert obtained from great distances. Whether the

Clovis dart or spear was launched by a spear-thrower (atlatl), thrown, or thrust, is not yet understood.²¹ Studies have shown that bone, ivory, and antler dart points, when sharpened, can do everything a chert point can do but without breaking. What was the purpose of such a difficult fluting flake? Other than Folsom, no one else in prehistory found such a feature necessary. Perhaps it was not about killing mammoths or large bison. After all, the Clovis kill sites do not indicate full use of the meat. Later sites reveal broken bones, fire-pits, and other signs of processing for storage. Why was the Clovis game at kill sites so lightly butchered?²²

Other than the ritualistic sharing of the easiest cuts of meat as status reinforcement, the megafauna kills may have been about male prestige, politics, or something like a Eucharist where the body of the deity is shared.²³ The earlier Clovis narrative holds that these people were specialized hunters who followed the megafauna on their migrations and killed them every time they had a chance. Recent archaeology describes the Clovis tradition as generalized hunters and gatherers, dependent on smaller game and vegetation. The big kill was not even a monthly occurrence; maybe it was only once a year. The megafauna would have been the subject of rockless campfire stories,²⁴ embedded in their creation myths and revered, not harassed. But there are a lot of mammoth and bison bones lying around the Clovis camps along with horses and smaller game, all of which were likely taken after rituals asking the animal's permission and according to protocol. In contemplating aboriginal motives, the best rule is to place symbolic meaning above the rational and utilitarian.

There is no evidence for mythology among people in deep history, just like there is no physical support for any behavior not revealed by a whisk broom in early soil horizons. But it's a sure thing that people told stories that became myths. People have always hungered for stories, and, as Joseph Campbell pointed out, the ancient shaman's function was to mythologize the environment and the world. Stories have the power to organize the world into a comforting, predictable place where good behavior is rewarded.²⁵ There may be evidence from Central Texas for Clovis-era storytelling— thin, limestone tablets, often shaped by percussion flaking, that contain markings. No one has a clue about these designs, and their interpretation is of course contentious; there are more than a dozen hypothesized meanings from simple graffiti to representations of time. Another possible interpretation is that these tablets were mnemonic devices, which were intended to aid the storyteller in recalling long narratives transmitted by song or chant.[26] The report of a Clovis point made of the legendary Alibates chert, found between two of these tablets, supports the theory of storytelling tablets.[27] The portable nature of these engraved stones and bones suggests the presence of a bard at hunting camps. Whatever they were, these incised stones accompany human archaeology back to 100,000 years ago when people were first energized to spill out of Africa.[28]

There is another class of incised stones studied at the Central Texas Gault site involving chert cobbles rather than thin tablets of limestone. These markings are difficult to study because they are in fragments. It appears that the raw Edwards chert cobble from the nearby quarry had been engraved on the soft cortex before the stone was processed through ritual and heat treatment. Unlike the limestone designs, the cobble markings were temporary, destroyed as the stone was reduced to its intended purpose. The purpose of the markings may have been divinations by a shaman who determined the stone's powers and destiny, placing a charm on the cobble to instruct its use. Then again, the incisions may have been only the owner's mark.[29]

Considering the evidence about Clovis and Pleistocene peoples leads to the question: What were they like? Bioarchaeologists, studying the bones of Pleistocene populations, find that these early people were taller than people today. Their teeth were not so crowded and lacked problems with the wisdom teeth; nutrition was better than after the Pleistocene, and interestingly, their skulls were more robust, long, and narrow.[30] Later Native Americans have a more delicate, rounder skull, which has caused some confusion and continuing assertions that there must have been any number of migrations from several regions. There is an explanation for why the Clovis and other Paleo-Indians do not resemble their progeny: they are the same people, but they evolved slightly over millennia. Archaeologist James C. Chatters, after studying the bones of people from 12,000 to 13,000 years ago, points out that more than half of the males suffered injuries caused by fighting, not by hunting. The healed skull fractures appear to be from domestic brawling rather than warfare. The much smaller women have different kinds of healed injuries—they have battered faces and suffered from malnutrition. Chatters's theory, known as the "wild-type hypothesis," explains that the hypermasculine males fought over the subordinate women, and these fights resulted in the development of these robust facial features of the earlier people. Then sometime after 13,000 years ago, things changed.[31] Hopefully, continuing research will mark the Clovis period as the time when the women began to select males with gentler facial features, reversing the adrenaline-charged, violent visage.

Historians look for interesting and imaginative explanations for past events, and it is interesting to speculate that the Clovis represented a return to a matriarchal society. But, alas, there are more compelling theories and studies on the equivalent era skulls among ancient Sudanese and Nubians of Northern Africa that offer more credible conclusions. Sex was not involved at all; the decreasing robusticity of the skull was a response to reduced mechanical demands on the chewing muscles. During the Pleistocene, these populations ate wild-sourced, hard-textured foods.[32] Texas archaeologists agree that the Paleo-peoples did not have metates to grind seeds, nor did they use rock ovens to process roots and tubers, and they did not leave much around their camps in the way of plant residue. If they used bison stomachs to boil meat to make it softer, there is just no evidence of it. So, it is a matter of Wolff's Law: bone tissue is added in areas where it is used and is taken away where it is not needed. When people stopped using their teeth as a third hand to tear fibers and chew so much, the teeth became smaller. So did the jaw, and a smaller mouth means a rounder skull vault and a less fierce visage.[33]

The Younger Dryas Event

Pleistocene cultures all over the world were on track to become larger, agriculture-based societies. Those incised rocks were soon to morph into clay representations and then into writing. But something happened in the Northern Hemisphere that pushed Middle Eastern societies forward while delaying agricultural development in North America. The historian is faced with hypothesizing between a dramatic event, the exciting stuff of science fiction, or something more mundane and credible. Whatever the cause, the Clovis hunting technique was no longer effective and faded out of favor, replaced by a new hunting technique. The world-changing event was the Younger Dryas episode that began abruptly 12,800 years ago, returning the Great Plains and Texas to more than a thousand years of cold, very dry weather.[34] There are not any archaeological sites along the Bosque River from this time period, yet more can be inferred about the period here than anywhere else in North America because of a thorough examination of sediment along Buttermilk Creek, near Austin, and Hall's Cave on the Edwards Plateau.

There is no consensus that explains what triggered this event, why the climate was so cold for so long, or why it ended around 11,700 years ago. Something happened to disrupt some combination of atmospheric or oceanic processes. The three leading hypotheses explaining the sudden onset are: (1) an airburst or impact by a comet, (2) a large volcanic eruption, (3) catastrophic drainage of a Canadian glacial lake. The more visceral explanation is that a cosmic impact, something like a large meteor or comet, exploded over or into a glacier. This is the scenario used to caffeinate class discussions. The evidence presented frequently describes a mysterious black mat, found at most of the Younger Dryas sites. This layer is variously said to be organically rich or consisting of microscopic diamonds, glass formed at high temperatures, carbon spherules, unusual amounts of platinum, and other things that indicate the presence of temperatures as high as 2,200 degrees Celsius.[35] For a few years the Hiawatha Crater in Greenland seemed to support the impact argument, but later research shows the crater to be too ancient. Most climate scientists have moved on, still searching for the trigger.[36]

The second area of research centers on the mega eruption of the German volcano, the Laacher See, which ejected nearly two tons of sulfur-rich particles into the stratosphere. This dust that reduced solar radiation and began the cooling of the Northern Hemisphere shows up in Central Texas archaeological sites.[37] But an explosive volcano, no matter how large, can't alter the climate for a thousand years. The third hypothesis contends that the warming trend at the close of the glacial period filled Lake Agassiz, a lake larger than all of the Great Lakes combined. This lake was caused by dammed-up chunks of ice breaking away from the melting glaciers. The timing of the collapse of the ice dam is a century or so earlier than the consensus of the 12,800 years ago onset, but such a massive flood could have initiated millennial changes. That much freshwater rushing into the Pacific Arctic or Atlantic Arctic Oceans or down the Mississippi River could have altered oceanic circulation for a very long time. Perhaps there was a shock to global atmospheric circulation or even a north-to-south air current alteration. What caused the Younger Dryas's climate reversal will remain an enigma until scientists sort it out.

The Folsom point originated on the Northern Plains around 12,800 BP and was in use until 12,200 BP or later. This thin point is recognizable by its full-length basal flute and careful flaking. The Folsom point-users specialized in hunting the *Bison antiquus*, which may explain why these points have standard-sized bases—to facilitate replacement on the hunt. These points are found mostly on the Western Plains from the Texas Gulf to Canada.

Briggs, Buchanan, et al., "An Assessment of Stone Weapon Tip Standardization during the Clovis-Folsom Transition in the Western United States," *American Antiquity* 83, no. 4, (2018); Briggs Buchanan et al., "Geometric Morphometric Analysis Support Incorporating the Goshen Point Type into Plainview," *History Faculty Publications*, no. 27 (2019); Michael J. O'Brian, "Setting the Stage: The Late Pleistocene Colonization of North America," *Quaternary* 2, no. 1 (2019). doi:10.3390. Illustrated by Jill Alford.

The Younger Dryas (12,800 to 11,700 years ago) began with a climatic shock that started the countdown to the end of the Pleistocene as well as the iconic megafauna that defined the period. The Central Texas ecology collapsed as sudden cold and aridity stripped away almost all of the trees and other vegetation, causing erosion that stripped the topsoil down to 25 cm. Mammoth ranges contracted as water and digestible C3 grasses became scarce.[38] There must have been remnant microenvironments where nutrition-stressed animals made last stands as hunters harried them to extinction. Large animals like mammoths and horses went first; the big-toothed cats and lions, ambush hunters, couldn't catch the Plains-dwelling, fast-moving bison, so they faded away soon after. With no carcasses to feed on, the bone-crunching dire wolves disappeared as well. The mammoth hunter's belief systems were shattered when their world no longer had meaning.[39] As mammoth hunting lost its status there was already a new hunting strategy demonstrating its effectiveness within the Clovis population.

Folsom Bison Hunters

The Folsom hunting technique originated somewhere on the western Great Plains around 12,800 years ago and soon, perhaps in just fifty years, became an established lifestyle. This was a new lithic technology and hunting strategy centered on the large *Bison antiquus* that flourished during the Younger Dryas cold and aridity. Other than the climate the best description of the Younger Dryas was the spread of drought-tolerant, bison-friendly, C4 grasses reaching from the Texas Coast to Canada.[40] Changing plant and animal dynamics was an opportunity for each group to take up the new lifestyle to originate a new set of creation myths and develop an operating mythology. These traveling hunters carried engraved bone beads that may have functioned as rosary-like story sequences and bone disks that looked an awful lot like chronometers.[41] The Folsom tradition followed the bison on their migrations for a couple of hundred years before their new lifeway dominated the western Great Plains and the West as born-again bison people.[42]

The highly mobile bison hunters developed a less bulky lithic technology. Instead of striking long flakes from a prepared core, the Folsom craftsmen reduced a spall of chert down to a preform, then prepared the preform to receive the center flute—a high-skill procedure that was probably done after knappers had presented their preforms, probably in a ritual context, to the flute-maker specialist. Finished Folsom points were too fragile to carry, so the final thinning flakes would be completed as needed by a moderately skilled hunter in the field.[43] The first consideration when hunting bison on the open Plains was to launch a dart from the longest possible distance. Atlatl darts have been reconstructed that can easily strike with great energy beyond a hundred yards.[44] Experimental archaeology has shown that Folsom darts penetrated deeply into bovine carcasses even when striking a rib. The piercing impact of the Folsom point was made possible by the ultrathin chert blade. The main consideration when away from a chert source was the reuse of the point. The points were hafted so that only the tip was likely to snap. The flute provided easy resharpening of a broken point several times without loss of thinness. As the points were resharpened, another innovation appeared—the sliding foreshaft mount that could be cut free, thus allowing the Folsom point to slide forward for the next use. When the group returned to the chert quarry, these used-up "slugs" were discarded, while new points were made.[45]

The atlatl is the dart-throwing device thought to have been in use since the Pleistocene. Greater velocity is achieved by extending the leverage of the throwing arm. Illustrated by Jill Alford.

Based on chert types, the Folsom hunters are thought to have had home ranges—the Northern Plains, the Rocky Mountains, and the Southern Plains–New Mexico–Southern Colorado, which included the Bosque River. The standardized Folsom design and recurrence of preferred chert types suggest that Folsom groups had a tighter trade network than did the more residential Clovis. Folsom hunters migrated around their home ranges, frequently meeting others to exchange quality cherts. There must have been trade fairs or some sort of gatherings with different regions to stir the gene pool, but whatever they traded it wasn't chert; preferred raw materials from one range are rarely found in another.[46]

After about four centuries, increasing populations of competitors began to share the Plains and the Folsom hunters faded away. The trend was toward Balkanization into separate breakaway traditions; new dart point types like Midland, Plainview (or were they Golondrinas?), Meserve, St. Mary's Hall, and others. All of these are found along the Bosque River, usually reworked many times in this chert-poor region of the Southern Plains. What drove the new demographics was the approaching end of the cold and aridity of the Younger Dryas around 11,700 years ago, the beginning of the Early Holocene. With the warmth and a little more moisture, vegetation started to return to Central Texas, especially cedar (juniper really) and live oak. All over North America vegetation spread out of remnant populations and resettled much of the Plains, creating diverse regions. Plant diversity returned after Younger Dryas, while the diversity in large mammals did not. Paleontologists have found that thirty-five genera of large mammals went extinct during the last of the glacial period. In Central Texas, it was mostly camels, helmeted muskox, mammoth, and caballine horses. In most cases, it was the loss of tolerable climate and vegetation that pushed them into extinction.[47]

3

The Early to Middle Archaic

At the beginning of the Early Archaic, 8,000 years ago, the Bosque River valley was covered with a tallgrass prairie with fewer trees. It was mostly cooler and a little wetter than in the early twenty-first century. The giant *Bison antiquus* had replaced the mammoth as the primary shaper of the landscape by selecting the cool, moisture-dependent C3 grasses and trimming the brush and trees. These cool-weather-loving bison were starting to migrate out of Texas as the less stable postglacial climate frequently warmed up a little too much. Aboriginal Texans stopped hunting the bison exclusively and began to hunt smaller game and collect and process plants for the first time on a large scale. There were new tools, like sandstone grinding stones such as the metate and mano, that were used to process plant bulbs baked in hot-rock ovens. As Archaic peoples no longer migrated great distances, they became more isolated and had to settle for less exotic cherts than those once traded and collected from hundreds of miles away. Isolation gave rise to new, regionalized, barbed dart point styles as corners were notched in a variety of distinctive types. Warming continued and life became more difficult as the Southern Plains continued to warm up.[1]

Hot-Rock Baking

The last Canadian dam of glacial debris collapsed 8,200 years ago, changing ocean currents and bringing a brief cool spell over North America. A century or so later Central Texas continued to become warmer and drier, reducing post oak trees and taller grasses; the Bosque River was becoming even less attractive to bison.[2] As these animals became rare, people

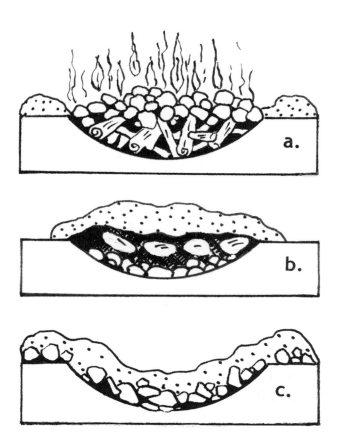

Known as burned rock middens, the earth oven site was reused for centuries. Each time the oven was prepared the fire-cracked rocks were removed, a fire was built, then the rocks were returned to cover the fire. Eventually, the iconic doughnut shape emerged.

After the fire burned down the stones were very hot; a thin layer of soil covered the heating elements. The bulbs, sometimes hundreds at a time, were encased in a layer of singed Opuntia cladodes or moist grass to provide steam. In order for the toxic plant bulbs to become edible the baking time was probably around thirty hours.

Hot-rock baking started with the Early Archaic around 9,000 BP and reached its most intensive use between AD 750 and 1400. Rock baking declined during droughts when the mostly lily family bulbs became scarce.

Johanna Hunziker, "Exploring Burned Rock Middens at Camp Bowie," *Texas Beyond History*, Center for Archeological Research, University of Texas at San Antonio (April 18, 2004). Illustrated by Jill Alford.

began to look more closely at their regional landscape for nutrition. Deer and smaller game provided some protein, but carbohydrates from roots and bulbs became the major food source in the Early Archaic. Harvesting roots and bulbs with a digging stick was labor-intensive, but to render these foods safe to eat, more nutritious, and digestible, they had to be baked for up to forty-two hours. Archaeologists interpret the so-called Carbohydrate Revolution as evidence of a growing population working hard to feed itself in a limited territory. This new foodway is known as "hot-rock baking."[3]

The oven was prepared by digging a pit a couple of feet deep and lining the bottom with rocks; a large fire was built on the rocks, and then the fire was covered with more rocks. After the rocks were heated, they were spread to cover the bottom of the hole. Moist, green vegetation—leaves, bluestem grass, or cactus pads with the spines burned off—was then spread over the hot rocks. The bulbs, or less often meat, were placed in woven bags or encased in grass bundles and laid on the hot rocks, then a layer of earth covered the oven. Archaeological experiments have shown that the rocks served as heat-energy reservoirs that slowly released the heat over forty-two to forty-eight hours. The baking process tenderized the meat and changed toxic plants rich in complex carbohydrates into softer, sweeter, calorie-rich foods. Residue on oven rocks reveals some starch-rich or inulin-rich favorites, like crow poison (*Nothoscordum bivalve*), winecup (*Callirhoe involucrata*), wild onion (*Allium canadense*), cattail (*Typha latifolia*), Indian turnip (*Pediomelum cuspidatum*), and the now rare wild hyacinth (*Camassia scilloides*). After baking, these roots and bulbs could be consumed right away, or placed on a sandstone metate, ground to a mush, and patted thin to be dried for winter use. There is no evidence from Spanish observers in Texas that this material was ever fashioned into tortillas, but it seems reasonable. Oven-baked foods would have been generally softer than Pleistocene foods and required less vigorous chewing. A few thousand years of rock-oven baked foods might explain the gradual transition to smaller teeth and the changes in cranial shape.[4]

It was during the Early Archaic that chert-knapping skills declined. After bison became rare along the Bosque River and only smaller game like deer and antelope were available, dart points (generally referred to as arrowheads) became a lower priority. As a general rule it seems that as plant food becomes more important, the quality of the projectile points becomes cruder. Preforms were no longer skillfully struck from a prepared core or

The Early Triangular point (9,000–8,000 BP) is found from north of the Bosque River to northern Mexico. The blade was sharpened by oblique flaking that eventually beveled the point. This projectile point—or possibly knife—was used during the Early Archaic period.

Ellen Sue Turner and Thomas R. Hester, *A Field Guide to Stone Artifacts of Texas Indians* (Houston: Gulf Publishing, 1999); and projectilepoints.net. Illustrated by Jill Alford

pressure-flaked into a thin dart point. The flake scars became shorter and less regular, producing a chunkier point. One of the first dart points of this class, though sometimes they were very thin, was the Early Triangular point that was produced over a very long time period, from between 9,000 and 8,000 years ago. And it might not have exclusively been a projectile point; many archaeologists think it may have been used as a knife. Only a few of these three-inch blades are reported in Bosque River collections, and most of those are broken or show impact fractures. But they do have signs of resharpening, sometimes resulting in a beveled blade. Some have careful flakes running upward from a concave base and have the appearance of a projectile point. It would have been difficult to haft one of these stemless points, and it would have detached in the wound, tearing flesh as the animal ran.[5]

The Gower point (8,000 to 6,000 years ago), less frequently found in Bosque River collections, is crudely made and distinguished by its deeply concaved base. The Gowers found on the Bosque River have been resharpened down to a nub and can be difficult to recognize, as is true with most points along the upper, chert-poor river. The same time period is shared by the Hoxie dart point, also crudely made, except it has a longer base that is less concave. Another Early Archaic type, the Uvalde, is recognized as similar to the last two dart points, except it has more pronounced barbs. The least elegant of the period is the Nolan point. It can be up to three inches long and almost barbless, the distinguishing feature, unique to this point

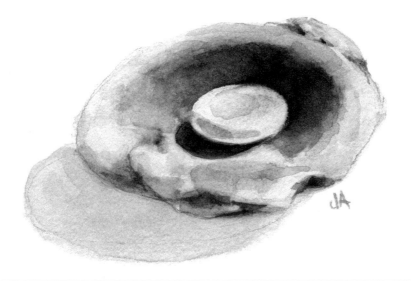

The manos and metates found along the Bosque River were most likely made from sandstone found in the northern part of Erath County. They were used to grind baked roots and tubers, acorns, and seeds. Those found in fields typically are scarred or broken by plows. Illustrated by Jill Alford.

type, is a beveled stem.[6] The Early and Middle Archaic are both known for the increase in woodworking tools, primarily the Clear Fork adze. The tool was also used in the Pleistocene when it was more carefully constructed. The Archaic types were thickly made, and use-wear studies suggest that these tools (found in both male and female graves) had a specialized use.[7] Perhaps they were used to shape atlatls and digging sticks.

The Altithermal (8,000 to 6,000 Years Ago)

This very hot period is known as the Middle Holocene Altithermal, when plant communities were reduced to "desert-Plains-grassland," and survival became more difficult for those who remained in Texas. The same mid-latitude atmospheric circulation changes that turned the Sahara green, heated and dried North America.[8] The moisture-loving C3 grasses and taller C4 grasses, like big bluestem (*Andropogon gerardii*), were replaced by tough warm-season C4 grasses. And except for the 200-year hot spot

Sideoats grama (*Bouteloua curtipendula*), a perennial Plains/prairie short grass, is found from Argentina to Canada growing in single bunches or clusters. Most of the nutrients are found in the six- to eight-inch-tall leaves. The seed stalks grow to two feet, with the seeds hanging from one side. Sideoats is the tallest of the grama grasses and has a wider tolerance of soil and is an early colonizer of drought or abusively grazed areas. Sideoats grama is the official state grass of Texas because of its economic importance. Botanical historians report that because the seed stalk resembles a lance, the Kiowa and other Natives wore it in their hair if they had killed a man by lance.

Matt Warnock Turner, *Remarkable Plants of Texas: Uncommon Accounts of Our Common Natives* (Austin: University of Texas Press, 2009); Frank W. Gould, *The Grasses of Texas* (College Station: Texas A&M University Press, 1975). Illustrated by Rainey Miller.

around 6,500 years ago, the most drought-resistant of these, like buffalograss (*Bouteloua dactyloides*) and blue grama (*Bouteloua gracilis*), increased. Studies from the droughts of the 1930s and 1950s have shown that little bluestem (*Schizachyrium scoparium*) and side oats grama (*Bouteloua curtipendula*) can endure longer into a drought than other fair-weather grasses but will give way during intensely arid conditions to a blue grama–buffalograss community. Buffalograss can spread quickly when the landscape opens up by having shallow roots that are ready to take advantage of rare summer showers, spreading an inch per day, locking down the soil so it can't blow away like it was doing in West Texas in the 1930s and 1950s.[9]

Pollen studies show that as most trees disappeared, the rugged post oak and even pecan trees held on. The climate was stuck in windy La Niña mode, which is characterized as persistently dry and marked with occasional heavy rains from tropical disturbances in the Gulf. The Brazos, and by inference, the Bosque River, experienced sediment stripping and lateral migration of their channels during the Altithermal. So much soil was eroded into these rivers that deep sediment accumulated downstream from Waco. The Texas shoreline was built up during these centuries of erosion. Thousands of years of Pleistocene and Early Archaic archaeology were scrubbed away with the banks of Texas rivers, which explains why these campsites are so rare.[10]

Opuntia (Texas Prickly Pear)

Considering how Texas prickly pear (*Opuntia engelmannii*, var. *lindheimeri*) is seen to rise to the occasion during modern droughty episodes along the Bosque River, it must have been a major food source during the Altithermal. Without evidence from dry rock shelters along the Bosque, the use of Opuntia can be inferred from the plentiful information from Trans-Pecos

Texas prickly pear (*Opuntia engelmannii*) has survived the worst of Texas droughts because of its ability to seal in moisture collected by shallow roots that benefit from brief showers. The entire plant is edible, including stems, flowers, seeds, and the tunas.
Matt Warnock Turner, *Remarkable Plants of Texas: Uncommon Accounts of Our Common Natives* (Austin: University of Texas Press, 2009). Illustrated by Jill Alford.

Texas. Opuntia and its many species descended from Cretaceous ancestry in South America; then some early forms migrated into North America during the warm part of the Pliocene epoch about four million years ago. Like the grasses, Opuntia expanded and contracted as it adapted to the Pleistocene's warming periods. Though not completely impervious to the driest conditions, the prickly pear evolved some drought protections; since leaves lose moisture, Opuntia leaves were reduced to spines, and Opuntia flattened their stems into energy-gathering pads called cladodes. They evolved skin that became thick and waxy to hold the water content at 90 percent. Like buffalograss and the gramas, the roots reach out widely and are shallow enough to catch mid-drought sprinkles.[11]

Every part of the Texas prickly pear was used for utility, medicine, and food. In the spring, when the first cladodes appear, they are spineless and can be eaten raw as nopalitos. These and the more fibrous summertime cladodes contain up to 8 percent protein (corn has 10 percent) as well as important vitamins. The mature cladodes contain bitter oxalic acid and must be baked or boiled. The purple fruit, tuna, ripens in late summer, and according to stories told by shipwrecked Cabeza de Vaca, the ripening and harvest of them was an exciting event. After rolling in hot coals to remove the tiny, barbed, hair-like spines, called *glochids*, the tunas were pounded to remove the sweet juice. After feasting, the remaining sugary pulp (70 to 80 percent sugar) was sun-dried into large "flatcakes" for winter use. Coprolite (dried human feces) studies from dry West Texas caves from this period contain more than 80 percent seeds from *tunas*. Ethnobotanists report that treatment of wounds was done by laterally splitting the cladode in half and applying the antibiotic, hemostatic mucilage side as a poultice to the injury or burn. The antibacterial property was put to use to clean questionable water by placing the pulp and gel over the surface of a gourd of water; after half an hour the substance sinks, leaving clear, drinkable water. An unusual use was to laterally split the cladode, sew it into a pouch to be used as a cooking pouch (one was found with a fish inside), or scrape away the pulp to create a cladode bag.[12]

Drought Foods

Pollen evidence establishes that the agave family member *Yucca arkansana*, found along the Bosque River, provided edible blooms in early summer, and the large asparagus-like shoots were roasted over coals for a special

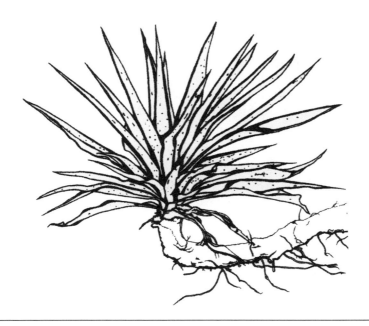

Yucca arkansana is a trunkless Agave composed of rosettes of long, narrow, sharp-pointed leaves. The blossoms are high in vitamin C and can be eaten raw or cooked. The fibers were processed to make fishing line, bowstrings, snares, baskets, and nets. Yucca cord was sometimes woven with strips of rabbit fur to make clothes and blankets. Yucca was an important soap, used as a detergent, body soap, and shampoo. Comanche ceremonies required people to shampoo their hair by shaving off slices of the root and pounding it into a foamy lather.
 Matt Warnock Turner, *Remarkable Plants of Texas: Uncommon Accounts of Our Common Natives* (Austin: University of Texas Press, 2009). Illustrated by Rainey Miller.

treat. The yucca blades were fashioned into cordage for jackrabbit nets, fishing lines, and sandals. Yucca, loaded with saponin, a caustic soap that is so strong that when the blades are scutched (pulp dissolved and the fibers removed) it is done underwater in which ashes have been dissolved, to protect the fingers. The saponin juice, leached from the crushed root, made a powerful lice-killing shampoo.[13] Texas thistle (*Cirsium texanum*) is another spiny food plant available in droughty times that can be baked or eaten raw. The leaves, stem, and taproot can be peeled and eaten with the crispness of a carrot. The seeds can be eaten or ground into flour for later use. And like so many foods in the near-desert diet, there was much chewing required.[14]

There were breaks in the Altithermal when the Bosque River must have flowed regularly for a decade or so, when deer and antelope were taken with atlatl darts, mostly tipped with the Nolan point. But if coprolite studies in southwestern Texas hold true locally, most protein was acquired from small mammals caught in yucca nets and snares. A favorite was the woodrat (*Neotoma*), historically reported to have been clubbed as they ran from their burning twig nest. Also popular on the menu was the cotton rat (*Sigmodon*), the cottontail rabbit (*Sylvilagus*), the drought-resistant jackrabbit (*Lepus*), the less frequently found pocket gopher (*Thomomys*), a couple of mice species (*Peromyscus*), quail (*Colinus*), ground squirrels (*Spermophilus*), and assorted lizards, fish, and birds, including "horny toads" (*Phrynosoma*). Nothing was wasted (and perhaps not shared) when these animals were eaten: for example, 47 percent of the coprolites in one cave contained rodent skull fragments, even teeth. Rabbit vertebrae and leg bones were consumed whole, while the smaller carcasses were softened by beating on rocks and eaten head first. Coprolites frequently contain hair, an indication that these small animals were not always cooked.[15]

If this sounds like a low-energy, high-risk diet, think how much worse it would be with parasites. There are well-preserved remains in coprolites of parasitic presence in the form of eggs and larvae of roundworms (*nematodes*), flukes (*trematodes*), tapeworms (*cestodes*), thorny-headed worms (*acanthocephalans*), and various genera of protozoan cysts. The parasites were restrained as long as people didn't camp in the same place for very long, but when drought-dwellers become more sedentary, close living probably doubled their parasite load.[16]

A Stressed Population

It's difficult to imagine the unsavory Altithermal foods, like insects, lizards, and the especially unpalatable "second harvest" reported from historic times. This second harvest occurred after the feasting of Opuntia tunas. The custom after the feast was to defecate in a designated location, then months later, when the band returned, pick out the seeds, perhaps cleaning them in antiseptic cactus pulp (hopefully), before metate-grinding them into an oil-rich flour for later use. It's a sure sign of stress when people are picking through their poop. Burials examined from 6,800 years ago in South-Central Texas show that Indigenous peoples suffered nutritional stress. The

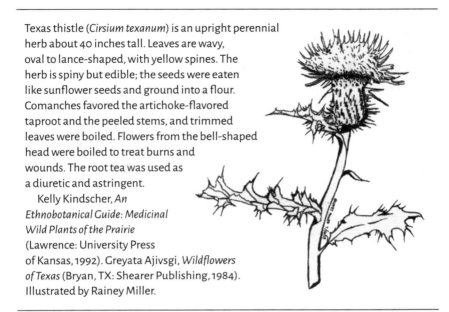

Texas thistle (*Cirsium texanum*) is an upright perennial herb about 40 inches tall. Leaves are wavy, oval to lance-shaped, with yellow spines. The herb is spiny but edible; the seeds were eaten like sunflower seeds and ground into a flour. Comanches favored the artichoke-flavored taproot and the peeled stems, and trimmed leaves were boiled. Flowers from the bell-shaped head were boiled to treat burns and wounds. The root tea was used as a diuretic and astringent.

Kelly Kindscher, *An Ethnobotanical Guide: Medicinal Wild Plants of the Prairie* (Lawrence: University Press of Kansas, 1992). Greyata Ajivsgi, *Wildflowers of Texas* (Bryan, TX: Shearer Publishing, 1984). Illustrated by Rainey Miller.

leading indicator is dental hypoplasia—defects in enamel caused by episodes of disease or lack of nutrition during childhood, while the teeth are still forming. The shock of sickness or near-starvation halts the production of enamel during the episode so that when the tooth erupts it has a line of missing enamel. Someone having a hard life may have several lines of hypoplasia. No other population of the same age in the Global History of Human Health Database has a higher rate than Altithermal Texas.[17]

One thing these desert foods have in common is their glycemic index, the measure of how fast a food absorbs its energy into the bloodstream. It's a scale of 1 to 100, with a doughnut close to 100. Foods scoring less than 55 are considered low on the glycemic index, and a meal of desert foods would have scored around 25. A characteristic of low glycemic foods is the amount of fiber; for example, the fiber of antidiabetic, near-desert foods was up to thirty times the amount of fiber in our modern diet. A few thousand years (longer for those in the Southwest for whom the Altithermal never ended) of this very low-calorie, high-fiber diet made the "Native American gut a paragon of efficiency."[18] The price for this near-perfect metabolism was the amount of fiber chewed, which indicated that they must have eaten all day to reach the estimated 200 and 300 grams of undigestible fiber ingested each day. Nutritionists today recommend 25 grams of fiber per day, and the average adult manages to get around an unhealthy 15 grams.[19] Teeth suffered more in the Southwest where the agaves were edible, but in Central Texas, where yucca is the only agave, the greatest offender was the Opuntia. As noted in the last chapter, plants contain glassy phytoliths, and droughty plants like agaves and Opuntia were especially loaded. Chewing these bits of glass removed enamel, sometimes all of it. In Southwestern Texas dental pathologists struggle to determine whether it was jagged phytoliths or metate grit that caused people to lose their molars as early as twenty-five years old.[20]

In South-Central Texas, and by inference, along the Bosque River, Native Americans during the Altithermal not only lost their teeth, but the accompanying periodontal disease caused severe health consequences. Constant dental distress and the characteristic abscesses and inflammation, which must have frequently co-occurred with pregnancy, are known to cause adverse health consequences in the offspring.[21] The Barker hypothesis says that early stress in life can damage the immune system and shorten life, as evidenced by several burials at the South-Central Texas Buckeye site. Competing with their parasites for nutrition, people during the Altithermal must have always been hungry, prickly pear pads rate only 7 on the 100-point glycemic index, and the ubiquitous enamel hypoplasia lines found in burials from this period show that sometimes there was just not anything to eat.[22] The unfortunate legacy of adjusting the digestive system to near famine, drought foods is a body that continues to respond as if starvation is a real possibility. Now that the ancient foods have been replaced with what some called the "pablum" diet of low-fiber, calorie-rich foods, the thrifty metabolism is left vulnerable to obesity and diabetes.[23]

4

The Middle Archaic

~~~~~~~~

The climate, warming for centuries, built toward a peak around 6,000 years ago. And then, the climate changed abruptly, bringing about a new cultural florescence that quickly spread over a half million square miles, with the Bosque River very nearly at the center. This was the largest territorial identity since Pleistocene times. The catalyst was a rapid and dramatic cooling event most likely triggered by differences in east–west temperature gradients or atmospheric and oceanic circulation changes.[1] The Southern Plains were cooler, but just as dry, and that was enough to attract bison. The grasses were already there, having spread during the megadrought; what had been missing was the cooler temperatures that bison preferred. Especially these bison, the soon-to-become extinct *Bison antiquus occidentalis*, a third larger than modern bison, that had also developed a taste for low-nutrition C_4 grasses before they arrived in Texas. The root-baking locals along the Bosque River had not seen bison in generations, and one wonders how many of them joined the exciting new cultural pattern that stretched from Mexico to Kansas, known by archaeologists as Calf Creek.[2]

The Calf Creek Hunters (6,000–5,750 Years Ago)

The American bison adjusted to a changing environment during which time adaptions greatly favored smaller bison. Large, long-horned bison from Eurasia arrived in North America around 130,000 years ago, and by the end of the Pleistocene, the early forms were already gone. The environmental shock of the Younger Dryas period caused a genetic bottleneck that ended the large Pleistocene fauna, and also the not-so-large mammals like camels and horses. Increasing hunting pressure along with the spread of heat-tolerant C4 grasses continued to result in downsizing throughout their history. The *Bison occidentalis* and the modern *Bison bison* survived this test because they had diversified into different ecological pockets; of these, the more successful adaptions featured the ability to live by grazing instead of browsing. They soon spread over the grassy Great Plains, becoming rare visitors to the warmest places, like Texas, until the climate cooled down 6,000 years ago.[3] The presence of bison changed everything.

It's probably not a coincidence that the two Altithermal period deer-hunting traditions, characterized by the Gower and the Uvalde dart points, disappeared from along the Bosque River at this same time. That development invites the question—were the Calf Creek bison specialists an invasion of advanced buffalo hunters that followed the beasts into Texas from wherever the bison had been during the Altithermals? Or, since there are no archaeological clues that support a violent intrusion, did the Calf Creek tradition rise from cultures all over the new bison range? If this replacement had occurred in Old World history, the expected scenario would have been an invasion of one ethnicity replacing another. But as yet there are no Calf Creek burials and so no DNA to support this notion. Archaeologists resist the whole idea of identifying users of a projectile type as a culture, so for now the Calf Creek point users were a successful bison hunting strategy that quickly caught on once the bison became available.[4]

When the *Bison occidentalis* extended their migration loops to include Texas, the Calf Creek hunting strategy and astonishing stone technology probably rose from existing cultures. The Calf Creek cultural pattern is older in Central Texas than in the rest of its range, and there are older broad-blade styles, like the Martindale (7,000 to 5,000), that could have been precursors.[5] Soon after the Calf Creek system came together around 6,000 years ago, it spread eastward and northward, following the bison migrations and demonstrating the new hunting technique to interested locals from Mexico to Kansas. Rather than a Calf Creek culture roving over the Southern Plains, these groups of bison people were more like a franchise in that to become a member, they had to learn the very difficult skills involved in the crafting of the Calf Creek series of atlatl-launched dart points. The skill level required to make one of these projectile points was never matched during the thousands of years of chert-knapping that followed.[6]

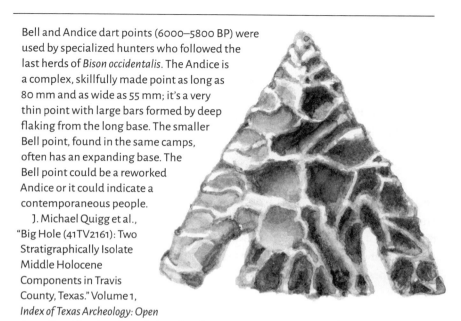

Bell and Andice dart points (6000–5800 BP) were used by specialized hunters who followed the last herds of *Bison occidentalis*. The Andice is a complex, skillfully made point as long as 80 mm and as wide as 55 mm; it's a very thin point with large bars formed by deep flaking from the long base. The smaller Bell point, found in the same camps, often has an expanding base. The Bell point could be a reworked Andice or it could indicate a contemporaneous people.

J. Michael Quigg et al., "Big Hole (41TV2161): Two Stratigraphically Isolate Middle Holocene Components in Travis County, Texas." Volume 1, *Index of Texas Archeology: Open Access Gray Literature from the Lone Star State* 2015, No. 1 (2016): 53; and Sergio J. Ayala, "Calf Creek Horizon Evidence at the Gault Site (41BL323): A Description of the Imagery Found in the Volume 5 Cover Border Design," *Index of Texas Archeology: Open Access Gray Literature from the Lone Star State*, 2019, no. 44 (2019): xi. Illustrated by Jill Alford.

The bison-supporting cool spell that enabled the Calf Creek phenomenon only lasted 250 years, which is a very short time in archaeology, rendering these points rare anywhere they are found. The two most common types in the series are the Bell (found along the Bosque River) and the even more rare Andice. Both types were reduced from large flakes struck from a larger piece of high-quality chert stone, which were then reduced through various flaking methods to thin, nearly hand-sized, triangular preforms. The preforms were buried and heated from above and below for twenty-four hours or so. The heat treatment renders the chert into a glassier material, which, when struck on the edge of the preform with a sharpened deer antler, will carry a flake all the way to the center of the piece, resulting in an extraordinarily thin template. The pièce de résistance that sets this point apart from all others is the notching technology. Each of the Calf Creek series of points is known for its very deep, parallel notches that create large, downward-facing barbs. The notching was done by striking a tiny, prepared platform on the base of the preform with a finely sharpened deer antler, which was struck by a mallet (indirect percussion). Each strike had to be perfect or the whole preform would shatter, and some of the finer points required up to sixty-five strikes. The nerve-racking efforts to replicate this procedure might lead one to wonder if the crafting of such a dart point was an application to join an elite bison-hunting society.

Close students of the manufacturing procedure have concluded that the craftsman did not start to make the top-of-the-line Andice point and then turn his/her efforts toward a Bell type after some mishap. The Bell and Andice templates and chain of operations were separate procedures. The simpler Bell process could have been learned by most hunters after a high level of discipline and practice. The Andice, however, was a work of art, requiring the dedication of the professional who probably did not hunt. Both of these points are found in the same camps, at the same time period, so it was not a matter of the skill improving or deteriorating over time. Archaeologists have determined that there had to be some "social/cultural reasons" for two skill levels of the same tradition existing side by side at the same historical moment.[7] Perhaps the Bell was the dart point used by the typical bison hunter, and the Andice point was awarded to those chosen to enter a hunting society, a high honor that must have included many perks not transmissible through history as easily as stone.

Life without Bison

Around 5,750 years ago, the Southern Plains became drier and warmed up again as changes in the Earth's orbit continued to increase the amount of solar radiation bearing down on the Southern Plains. No one knows what happened to the Calf Creek bison hunters. It is supposed that as the herds retreated northward, the hunters downgraded the classic Calf Creek series of dart points into less stylized yet similar points. Bison hunting stories and traditions were probably mythologized into legends that lasted a few generations. River sediment studies reveal that the climate began to stabilize into warm and dry conditions, punctuated by large, erosion-causing thunderstorms, much like today. Pollen and phytolith studies confirm that Central Texas was covered with short Plains grasses, with grassfire-limited stands of oaks, cedar (juniper), mesquite, walnut, pecan, elm, and mulberry.[8]

The large Calf Creek hunting bands were replaced by widely scattered, smaller foraging groups. Now isolated from the far-roving Calf Creek connections, these bands not only degraded the sophistication of their dart points, but they also simplified the manufacture by shifting from the more complicated biface (two-sided) to the one-sided uniface. This shift could imply that hunting became a lower priority. Even in these dry centuries, people's attention had returned to collecting seasonal plant foods, with an occasional protein boost from rabbits or deer. Rock ovens increased again, especially on the lower Bosque River, where the deeper, more fertile valley was a couple of miles wide. There were fewer rock ovens on the thin soils of the upper half of the Bosque River because the all-time favorite baking bulb, camas, has a low drought tolerance. Archaeologists believe that these self-sufficient bands came together seasonally, to stir the gene pool, compare knapping techniques, and collect and process whichever resource was abundant that year.[9]

The Mesquite Tree

During the warmest part of the Middle Archaic, about 5,750 to 5,000 years ago, there was at least one frequent source of nutrition. The honey mesquite tree (*Neltuma glandulosa*), whose seeds are found in Bosque River forage camps,[10] was probably controlled by fire to maintain a savanna-like population. Individual, fire-trained trees can grow straight, to around twenty-five feet tall, and can live to 200 years. The drought-tolerant mesquite is often a reliable crop when the other food trees failed to produce, and oddly the more intense the drought year, the heavier the yield. The mesquite produces tasty, fleshy pods that surround beans that can have more sugar than sugarcane, as much as 35 percent, and around 39 percent protein. A single tree can produce 20 pounds of pods if they are allowed to reach maturity. Much later, Spanish chroniclers reported that the flowers were collected and roasted in balls, and the green pods were used to produce a sweet syrup that was dried as candy. An amber-colored gum was harvested from incisions made in the bark that provided antiseptic gumdrops.[11]

The most important characteristic of mesquite was that the pods and even the meal can survive storage through winter. The tan to purplish pods were gathered in the fall. Since the sweetness and yield vary enormously from tree to tree, it's almost certain that leading families claimed

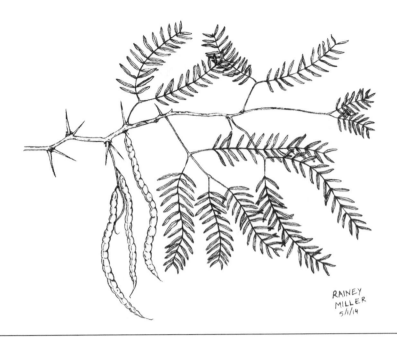

Mesquite, *Neltuma glandulosa*, once held in check by grassfires, is far more common along the Bosque River than it was centuries ago. Mesquite is a small tree, usually under forty feet, with an open crown and short trunk, unless trained by fire to grow taller. The nutritious fruit was once vitally important to Indigenous peoples, some of whom had names for the eight stages of pod development. The seeds contain about 39 percent protein; a long-lasting bread was baked from the meal.
Matt Warnock Turner, *Remarkable Plants of Texas: Uncommon Accounts of Our Common Natives* (Austin: University of Texas Press, 2009). Illustrated by Rainey Miller.

the fruit from heritage trees for generations. European observers noticed that because the process was so sticky, the roasted pods were pounded in cottonwood mortars rather than ground on metates, and the even more nutritious seeds were removed and processed separately. The meal was then stored or mixed with bone marrow and cooked as a porridge, boiled as "mezquitatole," or baked as a long-lasting flatbread. There was no pottery in the Middle Archaic, so any boiling would have taken place in an animal's stomach or stone-boiled in gourds. Later the Apaches fermented the sugary meal and water into a beer, but there is no evidence that Natives along the Bosque River brewed such a drink.[12]

The Oak Trees

After the mesquite bean pods had been processed, sometimes there was a fall bonus if acorns came into season. For millennia the most common oak tree along the Bosque River has been the post oak (*Quercus stellata*), preferring to grow outside the Bosque valley in sandy soil. The live oak (*Quercus virginiana*) grew in the dark soil formed by grasses on the hills and in the valley in the deep alluvial soils. The largest of the oaks, with the largest acorn, is the bur oak (*Quercus macrocarpa*), whose range reached into the Bosque valley only during times of increased moisture. The acorn crop depended on the masting cycle; typically, oak trees would bear a heavy crop once in three to five years. The masting strategy was intended to satiate the predatory insects and rodents, allowing several acorns to sprout. The collective signal is thought to be an average temperature and rainfall for some period in the spring, after which all of the oak species along the Bosque River were synchronized to boom or bust that year. There is no way to know, but if the heavy mast year takes its cue from the environment, then perhaps a close observer could make the same prediction. Shamanic prediction would have been especially difficult during the very warm and dry Middle Archaic because hot, droughty weather destroys the oak's flowering period. Then a shaman, noticing a heavy or light flowering, would be able to predict the acorn crop. Pollen studies show that during the warmest centuries between 5,750 and a little after 4,000 years ago, trees nearly disappeared in Central Texas,[13] which would have made a rare heavy masting year a cause for celebration.

When a big mast year occurred, far-ranging bands probably gathered to share the work of processing the acorns. As with other plant foods, the point was to store a valuable nutrition source through the winter. Acorn flour is known to contain around 8 percent protein, and up to 37 percent of much-needed fats, enough nutrition to make the difference in whether some families survived the winter. After collection, the acorns were dried, probably with the use of fire, before they were shelled and crushed. Texas observers from the nineteenth century were able to find the occasional post oak tree bearing acorns with low enough tannin levels to peel and eat raw. But as a rule, Texas oaks have high enough tannin content to wreck the kidneys and give acorns a bitter taste. To render the acorns palatable, the crushed pulp had to have been leached in baskets overnight in water. When

The roughly made Nolan point (6000–4000 BP) is identified by its beveled stem, weak barbs, and with random flaking. The point is most commonly found in Central Texas with the Bosque River near the center of its range.

Ellen Sue Turner and Thomas R. Hester, *A Field Guide to Stone Artifacts of Texas Indians* (Houston: Gulf Publishing, 1999). Illustrated by Jill Alford.

dried and ground, the meal was ready to store for winter use as flatbreads or in soups or stews.[14]

The Middle Archaic population increased along the lower Bosque River through the dry centuries, as evidenced by larger campsites. There were more burned rock middens, and, compared to the low ratio of dart points, there was a greater reliance on plant foods over deer hunting. The roughly made Nolan point (6,000–4,000 years ago) was randomly flaked without the pride of the full-time hunter. It originated as a long, up to three inches, barbless point usually found resharpened down to a barely recognizable nub. The one distinguishing feature is that the base is beveled. This slight twist could have been in sync with spiraled dart fletching, which could have produced a spinning projectile intended to keep the dart in a straight trajectory. Chert tools were generally cruder, often only flaked on one side instead of both, suggesting that hunting and leatherwork were secondary considerations.[15]

The harvested mesquite may have been one of the few semiarid foods during this period, available most years. The acorn crop was a rare treat, and the roots and bulbs for baking could not have been very reliable. The people experienced nutritional gaps evidenced by Harris lines on the long bones linked to periods of malnutrition, and in the teeth found in Middle Archaic burials, those hypoplasia marks on the teeth, discussed in the last chapter, remained at close to 50 percent during these dry centuries, evidence that very lean years were common. Dental caries were another unfortunate trade-off when carbohydrates replaced meat, almost doubling during this period. Throughout the Texas archaic, the teeth of both sexes show dental grooves and abraded surfaces, known from later observations to indicate the processing of fibers for cordage and basketry. Aside from the occasional acute episodes of nutritional stress shown generally in Central Texas burials, the archaic hunter-gatherers were in good condition and free of pathologies and infections.[16] And life was becoming easier during the last centuries of the Middle Archaic as atmospheric circulations began to deliver more rain and cooler temperatures after 5,000 years ago.[17]

5

The Wet Centuries

Beginning around 5,000 years ago the climate became more favorable to life along the Bosque River basin. Driven by slight changes in the redistribution of solar energy due to the earth's orbit, the next centuries were cooler, wetter, and stormier. Just as stronger El Niño years became more likely in North America, the same changes weakened the monsoon systems in Africa, and the Sahara turned into desert.[1] Pollen studies show that trees increased in Central Texas 5,000 years ago, providing a more abundant harvest and making the Bosque River more habitable. This ushered in a burgeoning population along with new occupation sites along the river. Several new types of dart points show up, indicating the growing population of the Bosque River valley was infused with new cultures from every direction. It was during these centuries that occupation sites contained the densest concentrations of cultural materials in the river's ancient history.[2] Perhaps the Bosque River valley became a little too crowded.

Plant Resources

The Bosque River basin was a greener place as rainfall increased. Millennia of eroded surfaces were covered with increasing vegetation as beaver dams filled deep gullies with sediment and moist meadows characterized the new ecology. Survival of spring oak blooms would have delivered larger mast years, perhaps once in three or four years rather than per decade.[3] Pecan trees (*Carya illinoinesis*) would have been able to spread away from the Bosque's floodplain, and though they mast on the same cautious scale as the oaks, they would have increased the food supply. Archaeologists think

Honey locust, *Gleditsia triacanthos*, is a thorny tree that can grow to 100 feet; the bean-like fruit pods ripen in the fall, growing as long as eighteen inches. The tree's seed dispersal strategy was shaped tens of millions of years ago by early ungulate mammals that were attracted to the 35 percent sugar-loaded pods. The tree is now rare because the seed is reluctant to germinate unless it passes through the digestive tract of horses or cattle. The pods wait for the Pleistocene pod-eaters, lingering on the tree or on the ground for up to three years. American Natives valued the sweet legume pulp as a food source and ground the seeds into flour; the flowers and leaves produce a tea with anti-inflammatory properties superior to aspirin.

Connie Barlow, *The Ghosts of Evolution: Nonsensical Fruit, Missing Partners, and Other Ecological Anachronisms* (New York: Basic Books, 2000). Chuan-rui Zhang et al., "Functional Food Property of Honey Locust (*Gleditsia triacanthos*) Flowers," *Journal of Functional Foods*, 18 (October 2015): 266–74; and Delena Tull, *A Practical Guide to Edible and Useful Plants* (Austin: Texas Monthly Press, 1978). Illustrated by Rainey Miller.

---

that the increasing importance of pecans may have been the deciding factor in locating winter camps. The pecan meats were far easier to process and more nutritious than the root crops. Pecans were rich in linoleic fatty acids, a nutrient that was much harder to find in ancient times. This high-energy food provided protein, carbohydrates, fats, and important vitamins. An

Southern cattail (*Typha domingensis*), a perennial aquatic herb, grows to nine feet, producing thick rhizomes with cigar-shaped flower spikes. The decay-resistant leaves are cured, then they are woven into mats, sandals, and plaited into baskets, rope, and twine. The flower heads are edible when young in late spring—raw or boiled and eaten like corn on the cob. Once the spikes ripen they burst open to release 200,000 nutty-flavored seeds that can be collected after the fluff is burned away. The fluff was used to pad cradleboards, diapers, and mattresses. The early shoots were cooked like asparagus, and the main food part is the rhizome, which is boiled, dried, and ground into a sweet-tasting flour for winter baking.
Matt Warnock Turner, *Remarkable Plants of Texas: Uncommon Accounts of Our Common Natives* (Austin: University of Texas Press, 2009). Illustrated by Rainey Miller.

important consideration was the low water content, allowing the nuts to be pounded together with dried venison and hackberries (*Celtus laevigata*), to produce long-lasting pemmican. A hot, creamy drink was prepared by boiling the crushed nuts, referred to in early accounts as pecan milk.[4]

The increasingly moist and fertile Bosque River valley was rich in food plants not available during the dry centuries. Honey locust (*Gleditsia triacanthos*) was creeping westward, with the help of white-tailed deer. The trees have two-inch-long spines used for leather punches, and the fifteen-inch-long, spiraled pods contain beans embedded in a sugary pulp (30 percent sugar in some trees), unlike acorns and pecans. The honey locust pods remain on the tree deep into winter. The young seeds, described as tasting like raw peas, were usually eaten when the whole pod was boiled. Pods processed at

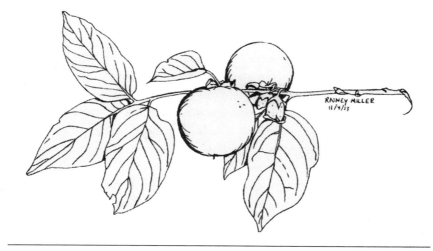

Persimmon (*Diospyros virginiana*) is a tree that is usually less than forty feet tall, with branches at right angles from the trunk; it is self-pruned as heavy fruit snaps the limbs. The dense, heavy, close-grained wood is a favorite for tool handles, gavels, and golf clubs. The inch-and-a-half-thick fruit is orange, astringent, and puckery to taste while green. The persimmon becomes sweet and luscious after the first freeze or overnight in the freezer, dissipating the tannins.

Comanches beat the fruit to a pulp, removing seeds and skins, then sun-dried it into a hard cake, to be used in winter cooking.

Matt Warnock Turner, *Remarkable Plants of Texas: Uncommon Accounts of Our Common Natives* (Austin: University of Texas Press, 2009). Illustrated by Rainey Miller.

---

an early stage were saved as a winter sweetener. Later, when the pods became bitter, the beans could be crushed and brewed as an antiseptic tea.[5]

An important wetland food source must have included the southern cattail (*Typha domingensis*), which was available all year, and its use must have been similar to later observed preparations. The most important part of the plant was the starchy rootstock, which was dried by roasting and pounding into a sweet flour that contained as much protein as corn. The flour was either stored for the winter or blended with processed acorns, mesquite, honey locust, persimmon pulp (*Diospryros virginiana*), or pecans to make shortbreads. In the spring the sprouts were available, and later in the spring the flower heads could be roasted and eaten right off the stalk. Cattail produces large amounts of protein-rich pollen, which was consumed as porridge or baked into bread. The presence of cattail pollen in human coprolites found in dry West Texas rock shelters confirms this gingerbread-flavored food. As observed in historic times, the leaves were woven into floor mats, sandals, and baskets, and, the mats covered the dwellings.[6]

Perhaps the closest the people along the Bosque River came to agriculture, thousands of years before it was practiced, was the maintenance of the blue-flowered camas (*Camassia scilloides*) meadows. Camas was one of the bulbs baked in rock ovens accumulating over millennia, referred to as middens. During these mesic centuries, hot rock baking increased as the root crops became prolific. Camas bulbs were the most important of these underground foods, and historians know something of their use from historical observations. Moist river-bottom meadows where these now-rare plants grew in abundance were maintained by burning to clear away any competition, such as the white-blooming death camas (*Zigadenus nuttallii*). When digging the camas bulbs, people learned to improve the area by replanting the smaller bulbs in soil that had been cleared of competing roots and inadvertently made richer by the soil mixing and addition of organic matter. The semicultivated bulbs, when ready in a few years, would have grown larger than the thumb-sized bulb known today.[7]

Animal Resources

On the prairies, there were pronghorn antelope (*Antilocapra americana*), but it was white-tailed deer (*Odocoileus virginianus*) bones that outnumber other animal species by far in archaeological sites along the Bosque River. Deer calories and protein are more nutritionally balanced than those from plants. As browsers rather than grazers, deer populations must have favored the loamy bottomlands where brushy growth provided the ideal habitat. The hides were used for clothing, the bones for awls, and the antlers for pressure-flaking tools for knapping tools and points. Venison was dried and stored. Additionally, spiral-fractured deer bones indicate that the equally valuable bone marrow and grease was rendered, and probably bagged for trade. Fish vertebrae were found in Bosque River encampments, so they must have been regularly eaten, but cooking must have hastened the decomposition of rare, delicate bones of turtles and reptiles.[8]

Not all food was about calories. Mollusks litter every Bosque River camp that was occupied for an extended period of time. There is little

Pronghorn, *Antilocapra americana*, is one of the few mammals to survive the Younger Dryas extinction event at the close of the Pleistocene (12.8K BP). Primitive cheetahs coevolved as the pronghorn's main predator, and, when the cheetah did not survive the Pleistocene this deer-like mammal, the only surviving member of the *Antilocapridae americana* family, was capable of running 60 mph, much faster than current predators. Eighteenth-century Spanish explorers called them "wild goats," and in 1844 while camping on the Bosque River, George Kendall wrote: "we saw an animal somewhat resembling both the deer and the goat, but with flesh preferable to that of either."

Walter W. Dalquest, "Mammals of the Pleistocene Slaton Local Fauna of Texas," *Southwestern Naturalist* 12, no. 1 (April 20, 1967): 1–30; Del Weniger, *The Explorers Texas: The Animals They Found*, vol. 2 (Austin: Eakin Press, 1997). Frederik V. Seerholm et al., "Rapid Range Shifts and Megafaunal Extinctions Associated with Late Pleistocene Climate Change," *Nature Communication* 11, no. 1 (2020):2770. https://doi.org/10.1038/s41467-020-16502-3. Illustrated by Jill Alford.

doubt that mussels were collected as a food source, but what about all those snails? The several species of gastropod land snails show up regularly in the screens of excavators as they investigate middens and have become a regular topic of conversation among both professional and avocational archaeologists. At contention is whether or not they were consumed as food by ancient Texas peoples. The most frequently encountered land snail found in Central Texas archaeological contexts is the one-inch *Rabdotus mooreanus*, a tall, white, conical shell with five to six whorls. The archaeologists who describe the *Rabdotus* as a food source point out that it is commonly found in midden material where cooking took place and, more convincingly, refer to the net bag of these shells found in a dig in Bexar County. Speculation has it that *Rabdotus* was gathered in a bundle of wet vegetation and placed on heated rocks for steaming.[9] The snails contain very little protein, and if they were eaten, they were more likely consumed for their vitamins.[10] To this date there is no actual evidence that the snails were eaten.

Few of the Bosque River archaeological investigations mention snails. But excavations all over Central Texas associate land snails with kitchen middens, so the question is, "How did they get there?" There is a growing consensus that the snails were attracted to areas of rich humus resulting from human activity. The snails were scavengers. An active camp was not an inviting environment, considering the foot traffic and earth-packing. While waiting for quieter times, snails can become inactive, plug themselves inside their shell, and go into metabolic arrest, an adaption they use to wait for rain. After the camp moved on to the next seasonal harvest, the midden area's composted refuse became an ideal snail habitat. When the archaeological excavator brushes the accumulated soil aside, they find dozens, sometimes hundreds, of mostly *Rabdotus* shells, as if they were on the menu.[11] Speculating about the snail as a food source is about as far as most investigators have been willing to go with this little mollusk.

One value of land snails to archaeologists is their sensitivity to environmental conditions. Identification of which snails thrived at different times can help reconstruct past environmental conditions. The sequence of land snails found in middens along Central Texas rivers confirms that the climate after 5,000 years ago was shifting from drier to wetter, leading into the Late Archaic beginning around 4,200 years ago. The prairie snail, *Rabdotus mooreanus*, an indicator of drier times, gives way to the similar *R. dealbatus*, and the flat, round *Oligyra orbiculata* that thrives in moist woodlands.[12] And these are only the larger snails caught in the archaeologist's screen, there are dozens more, like the tiny *Gastrocopta pellucida* and *Helicodiscus singleyanus*, that are far more sensitive to rainfall and climate change and pass through the standard quarter-inch screen. Sixty percent of the land snail species in Texas are micro-sized and can only be discovered through flotation devices.[13] Snails are a class of fine-tuned evidence yet to be exploited by investigators.

The other molluscan fauna whose food use is beyond question is the mussel, referred to locally as clams. Each of the four leading excavation reports from the Bosque River basin includes the presence of mussel shells associated with the Late Archaic period. The most frequently found mussel (69 percent) is the threeridge (*Amblema plicata*), which would have grown to seven inches if not for intense collecting pressure. The smaller minority species, never more than 10 percent, included the smooth pimpleback (*Quadrula houstonensis*), Louisiana fatmucket (*Lampsilis hydiana*), and a few others. All are threatened today because of a warming climate. Most of these shells found in middens are between three and four inches long; the valves are thicker than most and were favored for raw material for ornaments. These bivalves are identified by examining the umbo, the toughest part of the shell located just above the hinge, which typically preserves well.[14]

These mussels, supposed to have been steamed on a bed of vegetation on hot rocks, were nutritionally poor in terms of calories. It is thought that the mussels were consumed for their essential minerals like calcium and iron.[15]

Like snails, mussels have their habitat preferences; some like clear, running water, and others thrive in pools, like those formed by beaver dams. Mussels confirm an environment presented by other evidence, of a healthy river flowing through a forest gallery, with moist camas meadows and tributaries filtering through dense vegetation. This river ecosystem must have included plenty of fish, and some vertebrae confirm their presence, but cooked fish bones do not preserve well. The presence of fish is inferred from the types of mussels that can only survive in a habitat beneficial to fish.[16] Not long after the transition from the Middle Archaic to the Late Archaic 4,200 years ago, the locals were not able to keep the Bosque River basin to themselves.

Sharing the River

The increased population made possible by the improved climate and high-yield plant products led to the need for a more structured sociopolitical organization. Small, far-ranging groups were replaced by larger, almost sedentary base camps, needed to process the increased abundance. The settlement pattern of seasonal encampments in the spring and fall when family bands gathered to bulk process major food resources allowed the opportunity to stir the gene pool. Then afterward the base camp would disperse back into bands that followed the tributaries for a winter camp.[17]

The two basic types of camps were long- and short-term. The longer-occupation sites can be distinguished by the presence of burned rock middens, in which the extreme heat of several hours duration has heated the fire-cracked rocks into fist-sized chunks with right-angle breaks. Knapped chert scraps, scarce along the Bosque River corridor, were thoroughly used up, every flake left on the ground after someone's project was picked up and fashioned into a moment's-use tool. No large spalls or cores went unnoticed. The presence of manos and metates offers further clues that a camp was used extensively for a longer stay. A short-term hunting camp, more common farther up the Bosque River, can be recognized by burned rocks that are red but still in the original, rounded shape, which were used as heating or cooking elements in a low-temperature hunter's fire. Typically, the tools and points found in an overnight hunting camp are of the finest quality from downriver sources. If a chert core was found near a camp in a gravel bar, it would have been broken open to find single-use butchering flakes, with the large, valuable, unused portion left behind.[18]

Worldwide Climate Upset

The seasonally occupied base camp tradition was in place before the close of the Middle Archaic 4,200 years ago. This was when there was a worldwide climatic shock that set in motion the disruption and dispersal of societies. The mid-latitudes across the Northern Hemisphere experienced abrupt megadroughts that lasted a hundred years. The central authority in complex agricultural societies disintegrated and people began to move. In Mesopotamia, the Akkadian Empire collapsed, as did the Old Kingdom in Egypt, Liangzhu culture in China, and the Indus Valley civilization. All within a few dozen years. The cause is still being studied, but a change in solar irradiance is suspected. In North America, the megadrought was centered mid-continent, sparing Central Texas from catastrophic collapse of the region's ecology. It was still frequently droughty in Texas, but generally, these were wetter, cooler centuries. Along the entire Brazos River drainage, which includes the Bosque River, the floodplains continued to be freshened and built up by plentiful rainfall and flooding events.[19]

The sudden reversal of climate and ecological conditions 4,200 years ago was important enough for the Union of Geological Sciences, in July 2018, to name the third and current stage of the Holocene the Meghalayan Stage, to begin 4,200 years ago.[20] This date also coincides with the beginning of the Texas Late Archaic, when the Bosque River valley was shared by several ethnicities that were attracted to the river's resources. From the northeast, people who were making the Carrolton, Dawson, and Kent dart point styles shared the area with those making existing Central Texas styles, like the thick, heavy Bulverde and the more delicate Ellis.²¹ More people meant more competition for resources, which led to a sense of territory. Large cemeteries appearing along the lower half of the Bosque River confirm archaeologists' conclusions that people had developed a sense of belonging.²²

Bison Return to Texas

Arriving from the south around 3,600 years ago were the people who made the Pedernales dart point (3,600 to 3,200 years ago). Every collector will agree this is the most common Archaic point type found along the river. Compared to the duration of most dart point styles, four centuries was a brief appearance. These Indians occupied Central Texas during the centuries before bison again extended their range into this area. The Pedernales-associated people were part of the community that maintained a root-foraging, hot-rock cooking, and deer-hunting way of life. The Pedernales points faded from use during the brief time that cooler weather enabled the reintroduction of bison, now the 40 percent smaller modern species.²³ The reason for the cooling is not yet resolved but is thought to

---

Pedernales (4000 BP–3200 BP) is a generally triangular dart point that varies over its range from the Bosque River to the lower Rio Grande Valley. Before breaking and resharpening, the points are typically 3.5 inches long. This specimen has been reworked several times. Note the typical small flute as the base. Microscopic analysis of the edge shows wear consistent with use as a butchering knife as well as a projectile point.

Kristine Fischer, "Form and Function: A Case Study Using Pedernales Points from the Gault Site (41BL323) in Central Texas," master's thesis, University of Exeter, 2015; Ellen Sue Turner and Thomas R. Hester, *A Field Guide to Stone Artifacts of Texas Indians* (Houston: Gulf Publishing, 1999). Illustrated by Jill Alford.

The Wet Centuries | 53

The Castroville dart point (3000–2400 BP) was the largest of the broad-bladed bison hunting points from the Late Archaic, sometimes as long as 100 mm. Unlike the similar Marcos point, the Castroville is notched from the base to form the large barbs. Like many Bosque River dart points along the chert poor Bosque River, the Castrovilles are typically found resharpened several times down to a barely recognizable remnant.

Jon Lohse et al., "Toward an Improved Archaic Radiocarbon Chronology for Central Texas," *Bulletin of the Texas Archeological Society* 85 (2014): 274–78; Ellen Sue Turner and Thomas R. Hester, *A Field Guide to Stone Artifacts of Texas Indians* (Houston: Gulf Publishing, 1999). Illustrated by Jill Alford.

have been caused by another round of volcanic eruptions.[24] Radiocarbon dates are not plentiful enough to determine whether or not these hot-rock bakers turned to hunting buffalo. Texas was only bison-friendly from 3,290 to 3,130 years ago. During that time, the broad, long-barbed, four-inch, thin-bladed, bison-killing Marshall style replaced the Pedernales and is generally thought to have been the bison hunting point for this period.[25]

After four warm centuries, the hot-rock bakers along the Bosque River again redefined themselves as bison hunters and shared the river with outside hunters. The Homeric Grand Solar Minimum, which reset demographics around the world,[26] brought cooler weather that pulled the herds into the Southern Plains again around 2,700 years ago. This time Texas Indians, along with Plains peoples, hunted bison for an estimated 550 years. The two bison-associated dart points that date to this period are the broad, thin, heavily barbed Marcos and Castroville types.[27] During the middle of this period of bison abundance, around 2,500 years ago, evidence in the flood sediments of the Brazos River system shows a series of tremendous rain events. As part of this system, the Bosque must have shared the

The giant, long-horned steppe bison (*Bison priscus*), arrived in North America across Beringia about 230,000 years ago. Around 22,000 years ago this large bison was replaced by the slightly smaller *Bison latifrons*, which was replaced by *Bison antiquus* during the climate upset at the close of the Pleistocene around 10,000 years ago. As the Great Plains warmed, the grassland composition became less palatable to this ancient bison, prompting it to give rise to a transition stage (*Bison antiquus occidentalis*). During the cold, dry centuries after about 6,000 years ago, this last giant bison was replaced by the smaller, modern American Plains Bison, (*Bison bison*), shown here.

Jon C. Lohse, Marjorie A. Duncan, et al., *The Calf Creek Horizon: A Mid-Holocene Hunter-Gatherer Adaption in the Central and Southern Plains of North America* (College Station: Texas A&M University Press, 2021); Bjorn Kurten, *Before the Indians* (New York: Columbia University Press, 1988). Illustrated by Jill Alford.

channel-changing violence shown by the greatest river reorganization in 6,000 years.[28] There is a nearly six-foot layer of flood sediment at a lower Bosque River occupation site that may date from this period. And then the river's structure stabilized,[29] renewing the fertility of the plain.

By 2,150 years ago, the bison had receded northward again, and a thousand years of plentiful rainfall encouraged good root harvest. If the reports

The Wet Centuries | 55

by the Spanish in the 1600s were accurate, there must have been dozens of ethnicities in a Central Texas socioeconomic network maintained by coalitions and alliances that traded and shared information about changing resources.[30] Dental evidence suggests that past contact with East Texas and bison hunting incursions did not alter the Central Texas gene pool, at least not until later catastrophes.[31]

6

Social Disruption and Change

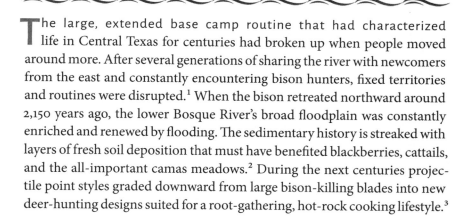

The large, extended base camp routine that had characterized life in Central Texas for centuries had broken up when people moved around more. After several generations of sharing the river with newcomers from the east and constantly encountering bison hunters, fixed territories and routines were disrupted.[1] When the bison retreated northward around 2,150 years ago, the lower Bosque River's broad floodplain was constantly enriched and renewed by flooding. The sedimentary history is streaked with layers of fresh soil deposition that must have benefited blackberries, cattails, and the all-important camas meadows.[2] During the next centuries projectile point styles graded downward from large bison-killing blades into new deer-hunting designs suited for a root-gathering, hot-rock cooking lifestyle.[3]

The Crowded River

As the broad-bladed bison-hunting points like Castroville and Marshall faded from use, they were replaced by smaller, less heavily barbed styles like the Darl and the Ensor (AD 200 until after AD 600). Ensor is probably the most commonly found dart point along the Bosque River, where it possibly originated, and spread over Central Texas. Biologists say that a good clue as to a plant's origin is an area with the greatest number of related species, like wheat in Afghanistan, corn in southern Mexico, or pecans in Texas. The many Ensor variants along the Bosque River illustrate this principle. An examination of the Ensor-like points in collections from along the Bosque River suggests that the Ensor is among the several independent bison to deer adaptions.[4] The Ensor coalesced into a dart point type along

The Ensor Dart Point (2200 BP–1400 BP) is the only projectile point style thought to have originated along the Bosque River, perhaps in the Waco area. The Ensor appears at the beginning of a period of bison scarcity. The type is characterized by a triangular, corner-notched, wide-based medium point between 30 mm and 70 mm in length. The bases in Bosque River collections range from slightly concave to slightly convex. There is evidence that when the use of the bow reached the area in the AD 600s, this dart point evolved into the Scallorn point as it was made in miniature to accommodate the smaller arrow shaft.

"Waco Lake: Bosque River Basin and Beyond," *Texas Beyond History* (September 2009); projectilepoints.net; Ellen Sue Turner and Thomas R. Hester, *A Field Guide to Stone Artifacts of Texas Indians* (Houston: Gulf Publishing, 1999). Abigail Peyton et al., *Data Recovery Investigations on the Eastern Side of the Siren Site (41WM1126)* (Austin: SWCA Environmental Consultants, 2013); Jon Lohse et al., "Toward an Improved Archaic Radiocarbon Chronology for Central Texas," *Bulletin of the Texas Archeological Society* 85 (2014): 274–78. Illustrated by Jill Alford.

---

the Bosque River a couple of hundred years earlier than the Ensor points found in the rest of Central Texas,[5] supporting the idea that this identity originated along this river. The Ensor style remained the marker point associated with hot-rock baking and population increase during the next eight hundred years.[6]

The Ensor point users were prominent along the Bosque River, which they shared with several other peoples. The river's resources were seasonal and often, like the acorn crop and pecan crops, not available every year. There was no part of the river so valuable that a group could enclose and defend it. In the centuries before agriculture Texas Indians had to rotate seasonally through their undefended territory in agreement with other groups. Survival and some expectation of security depended on participation in an information-sharing network. Shamans, or some such people, may have studied the area and arranged sophisticated scheduling for the different bands, perhaps detailing which group had access to which camas meadow that year and how many deer were to be harvested from which sector. Archaeological investigations into Bosque River sites show that different dart point identities camped in the same places, but probably at different times of the year. There must have been agreements, along with avoidance, that kept the peace most of the time.[7]

Most Texas archaeologists will resist connecting ethnicity to a projectile point type. Sometimes archaeology presents intriguing evidence to support the identification of a people from their lithic style. The Ensor identity was associated with Central Texas hot-rock bakers, they were the Bosque's home group. One of the other groups that frequently harvested the area's resources was the Godley point makers, believed to have been intruders that originated northeast of the Bosque valley. There is no way to know how these two groups got along with each other even though they shared the same territory. There is a Bosque River excavation in which the Ensor and Godley point users selected different colors of chert for their points. The chert in that area was obtained from the same river gravels in which both colors are available. The Ensor people consistently used dark gray or black, while the Godley outsiders chose pale gray. The displaying of colors as an identity marker is pretty basic human behavior, the sort of thing that must have led to violence more than once. There is at least one Central Texas burial with an Ensor buried in its ribs, as an indicator of violence.[8]

There was also a Darl point embedded in that skeleton, suggesting that the two peoples were allied. The Darl dart point has the distinction of being in use for centuries after the bow came into use. Around this time, contact with the north and east diminished, and the Central Texas cultures looked inward until the shock caused by the next time that bison flooded into Texas. That would not happen for another thousand years.[9] Long before the bison came back, the Godley, Ensor, and Darl point styles would fade from the historical record for a few centuries during a worldwide catastrophe.

The Dark Ages (AD 536 to AD 700)

Even though there were some dark years during this time, the term *Dark Ages* has fallen from use. In the literate parts of the world, there is plenty of information about what it was like to live through the Dark Ages, but on this continent, the term *dark* also refers to the archaeological record. Several of the consistently used campsites along the Brazos valley that have an ongoing record of occupation in the sediment have a sterile zone dating to about this time. People were not camping in their usual places until after AD 700. There is no evidence of a catastrophic population collapse, though that may be a possibility. A more likely scenario, considering the way Texas Natives responded to smallpox and cholera in historic times, is that the people fragmented into family groups and avoided the usual camps while they dealt with the epidemic. The only thing archaeologists can agree on is that there is a gap in the record after which the bow technology arrived in Texas.[10]

Several unfortunate events contributed to the social upheaval during this period. The first was a cooling caused by a cluster of large volcanic eruptions in AD 536, 540, and 547, made worse by disrupted ocean and sea-ice feedbacks, and a solar minimum.[11] The eruptions led to cold summers recorded in the biological memory of tree rings. Drought, food-gathering failures, disease, mass migrations, and famine occurred across the Northern Hemisphere from China to the Mediterranean to the Americas.[12] The first of these eruptions (AD 536) was Ilopango, near San Salvador, a caldera series with a long history of large-magnitude explosive volcanism. The ash-cloud fallout from this eruption covered the southeast Mayan highlands (fifty miles away) with knee-deep pumice and ash and loaded the atmosphere with sulfuric acid aerosols that covered the Northern Hemisphere for eighteen months. And when reinforced by the follow-up eruptions, the event was large enough to leave a sulfuric acid signature in both Greenland and Antarctica.[13]

The aerosol veil reduced the sun's intensity enough to cause summer frost from AD 536 through the 540s. Ice cores, tree rings, and ocean sediments confirm that this decade was the coldest in the last 2,000 years.[14] It was cold enough to have ruined crops all across the Northern Hemisphere. An eyewitness account from the East Roman world wrote: "It came about during this year [536–537] that a most dread portent took place. For the Sun gave forth its light without brightness, like the Moon, ... And from the time that this thing happened men were free neither from war nor pestilence nor any other thing leading to death." Another historian, writing from Mesopotamia, noted that "the sun began to darken ... there was distress ... among men ... from evil things." And an Irish monk wrote in the *Annals of Ulster* that there was a "failure of bread." A Chinese author recorded that " yellow dust rained down like snow" and described bitter frost that ruined crops.[15] The Bosque River valley analog must have seen a similar catastrophe; the camas and other root crops surely failed, while equally important pecan, acorn, and mesquite harvests were absent for years.

The beginning of the Dark Ages was more than unsettling visuals and a long cold spell, something happened to the landscape when rainwater collected the dust sulfates and reached the ground as caustic acid rain. There is a hyperbolic Welsh description of this rain as "vermin falling from the air, like moles with two teeth, devouring everything."[16] A sixth-century monastic writer described large-scale and widespread decimation of the landscape by fires; reports from China and Japan described fields made infertile.[17] The sudden toxic change in the soil chemistry for anyone depending on the earth for their sustenance, like the people living along the Bosque River at this time, would have been catastrophic. For a time, the river's water must have been toxic enough to kill fish and sicken the animals, the Indians would have had to move immediately to springs to find potable water.

The first of the Four Horsemen to reach Texas and the rest of the Northern Hemisphere was the black one, famine. The term *Dark Ages* refers to something besides the Kafkaesque gloom, another often inferred characteristic is an aversion to rational behavior. A cause for this behavior could have been aluminum, an element released by acid rain into the water and soil that would have lasted for years. Aluminum has devasting effects on the mind. Other things released into the soil by acid rain, cadmium, and lead, would have left the brain even more defenseless against the neurodegenerative properties of aluminum. Numerous studies have shown that aluminum causes nerve cell degeneration and inflammation in the brain, producing dementia, which must have contributed to the "irrational and extreme forms of collective behavior" that characterized the Dark Ages.[18]

The misery of the Dark Ages was compounded by another natural event that added disease to starved, irrational populations. Though it seems contradictory, grand solar minima are associated with pandemics. The last seven solar minima coincide with the last seven pandemics listed in the historical register of disease outbreaks. The mechanism that gives rise to new disease mutations is an increase in the flow of thermal neutrons, which mostly happens during periods of lower solar activity. There was a grand minimum of solar activity at this time, known as the Roman Minimum;[19] plagues were beginning in AD 541.[20] Each continent had an outbreak of disease, but the one most discussed in Western history was the Plague of Justinian beginning in 541 to 543 and recurring for a couple of centuries thereafter. The summer frosts and acid rain had caused crop failures in the eastern Roman Empire; in response, Emperor Justinian ordered grain—and the plague from Egypt.[21]

This happened as the cool weather caused a boom in the Oriental rat flea population, *Xenopsylla cheopis*, which was then jumping from the Nile rat, *Arvicanthus niloticus*, to the black rat, *Rattus rattus*. The flea/rat combination picked up the African *Yersina pestis* (bubonic plague) bacteria and arrived on the grain ships to Constantinople in 541.[22] This was the first time that this now-extinct form of this ancient bacteria evolved to super-spreader status. Transmission was mostly by infected flea bite, but when people crowded together in an effort to isolate themselves from others, droplet inhalation proved just as deadly. Nearly half of the city died, and when the plague spread across Europe, an additional twenty-five million died. That was just Europe, there were pandemics around the world.[23] The history of human societies everywhere was changed by migrations, political/religious reorganization, and violence.[24]

The Americas were spared from the devasting epidemics that drove social transformation in the rest of the world—or maybe the rest of the world was spared from the epidemic that swept North America. There was an endemic disease working in the background of Aboriginal American history, waiting to be released by the same climatic conditions that plagued the rest of the Northern Hemisphere. The disease caused by *treponema pallidum*, an early ancestor of syphilis that was nonvenereal, chronic but not fatal, and had been around the Americas for millennia.[25] The damage done by this disease wrecked a person's appearance far worse than anything brewing in the rest of the world. The infection was so intense that it left its signature on the skeletons, revealing massive bone-changing cranial sores, collapsed nasal apertures, bony swellings of the jaw, and leg bone deformity. Some forms de-pigmented the skin, which caused the victim to have an otherworldly appearance. The disease had been working its way through the more populated areas of the Americas, where some resistance was developed, but in Texas the epidemic could have been exaggerated; with a long history of hunting-gathering in small groups, the Bosque River valley had been left out of the bacterial arms race.[26]

A theoretical narrative would place the point of origin of a new treponemal strain developing in the Mississippi River valley, perhaps in the last decades of the Hopewell Culture. Corn-based villages were new and experimental in the AD 300–400s during the Middle Woodland period in eastern North America. There were constant introductions of corn from the southwest until a strain developed genetically to endure colder conditions. There is evidence that occasional crops were added to the usual plant-gathering as people coalesced into villages.[27] But they were getting ahead of themselves, using corn before beans, and the nutritionally balanced Three Sisters combination (corn, beans, squash), even as a supplementary source, would have left a population iron-poor. Along with the crowding that accompanied collective experiments in agriculture, people became susceptible to crowd disease. Especially if a solar minimum caused a jump to a new form of treponematosis.[28] The village experiment failed with the sixth-century climate catastrophe, and whether or not a disease outbreak was part of the social fragmentation that followed is compelling, yet still speculative.[29] The social disintegration that marked the decades after AD 536 could have been caused exclusively by the unusual number of volcanoes and the Roman Solar Minimum.[30]

East of Texas, during the Late Woodland period (AD 536–1000), villages and intensive food processing campsites were abandoned for a couple of centuries. The period was archaeologically dark in Texas, but there were more people in the east and so they have left more clues. There was a pause in outside trade and ceremonial burials and mound-building, the population dispersed, and some of the eastern family groups fortified their camps.[31] Stories told about the horrors associated with the Central Texas camps were probably responsible for the centuries-long absence at these sites. When the archaeological veil lifted around AD 700, the earlier hunter-gatherer lifestyle continued along the Bosque River. The most visible innovation was the adaption of bow and arrow technology, a more aggressive, rapid-fire weapon that enabled small groups to defend themselves.[32]

The Arrival of the Bow Technology

The use of the bow and arrow diffused southward from the Arctic where it had a centuries-long history, arriving in Texas during the unstable decades around AD 600.[33] Texas archaeologists are still struggling to build an explanatory narrative, because the new technology appeared in Texas so quickly that the transition left hardly a trace in the already vague archaeology during this period. A work of history can speculate cautiously from consensus information to reconstruct the story: there was a drop in population in Central Texas, people abandoned larger groups to roam in smaller bands, cemeteries and territories were abandoned, trade ceased, and there was a spike in violence in which the Scallorn arrowpoint was the cause of death.[34] This information combined with the treponematosis outbreak that left its signature on the bones of the early Mississippian corn farmers who mixed disease strains while crowded into their farmhouses to escape the summer frosts, which produced the pandemic in the river valleys that was spread southward, toward Texas, by desperate refugees. As one group collided with another, the use of the bow empowered small family groups to defend themselves.[35]

There is no reason to believe that the migrants themselves reached Texas in large numbers; a more likely scenario is that the social disruption caused by the pandemic produced chaos as the horror-struck populations overran time-honored boundaries, resulting in violence in which the bow-users prevailed over the atlatl users (spear throwers). The new bow technology would have made a terrific first impression as people saw a handful of archers put a larger community to rout. The atlatl had been in use for millennia, and it had its advantages; for one thing, when the five-foot-long dart was launched, it struck with greater kinetic energy than an arrow and might have been superior in bringing down large game. But there were plenty of disadvantages in conflict: to launch a dart, the thrower had to reveal his location by raising the throwing arm; the reload rate was slow; and the atlatl user probably carried only a few shafts.[36]

Faced with a single archer armed with a quiver full of arrows who could have released a dozen arrows in a minute or less, the atlatl user would have conceded territory every time. Atlatls were most likely ambush weapons, while the bow could have been used in offensive raids and more proactive hunting. The moment that a new group acquired the bow weaponry is open to theorizing; perhaps the treponemal disease affected mature adults more than young people, then technology would have surged as near-children were more open to new technology. Archaeology shows that there were plenty of people who ignored the bow and continued to use the atlatl for centuries. The same people likely used the bow and atlatl for different purposes—or perhaps the atlatl users were the old guys refusing to learn the new technology.[37]

There is one clue: perhaps the remnant atlatl users lived in deference to the bow people—the dart points began to shrink. It's possible that the mostly Darl and Ensor point users along the Bosque River began making smaller points to gain a few extra yards of dart flight, better to compete with the bow. Perhaps the atlatl users were just part of the group that found the dart superior in some types of hunting. Another explanation might be that the people with atlatls were confined to less productive territories without fine cherts or access by trade to the traditional quarries once used by their ancestors. With limited movement, dart points would have to have been made from whatever chert cobbles rolled down the river after the last flood.[38] Unfortunately, most of the Bosque River's gravel bars contain hardly any cobbles larger than a small fist.[39] Projectile points and other bifaces knapped from these small pieces would have been smaller, and smaller, as they were resharpened from use and breakage.

After AD 700 life along the Bosque River settled back into the Archaic lifeway, now with fewer people and smaller foraging and collecting parties. Aboriginal Texans who had acquired the bow continued to tolerate the dart users among them and life continued, though without the broad exchange networks available before the catastrophic sixth century. There was a slight degradation in lithic skill as the arrowpoint could be made from just a flake of chert struck from a larger stone. Before long no one bothered to go through the Archaic stages of production. This is not to say that arrowpoints were crude; the flakes were unsystematically reduced from the chert cobble, then finely pressure-flaked with a sharpened deer antler into very thin arrowpoints. Another convenience was that ancient quarries along the tributaries of the lower Bosque had been picked clean of larger cobbles, and spalls could now be harvested for leftover fragments suited for arrow-making.[40]

There are no known rock shelters along the Bosque River with dry enough layers of soil in the floor to preserve examples of the arrow shaft. There are several examples of shafts in West Texas and one nearby on the Brazos River near Blum, Texas. Like the longer atlatl dart shaft, the arrow was constructed of river cane (*Arundinaria gigantea*), often with a tough dogwood (*Cornus drummondii*) foreshaft, fitted into the forward end of the arrow. The arrowpoint was hafted onto the foreshaft so that, on the penetration of the animal, the forward piece would unplug rather than be ruined as the animal thrashed through the brush. The nock end of the cane arrow was also fitted with a hardwood plug to keep the fragile cane from splitting on release from the bow.[41] The arrow was fletched with woodpecker, buzzard,

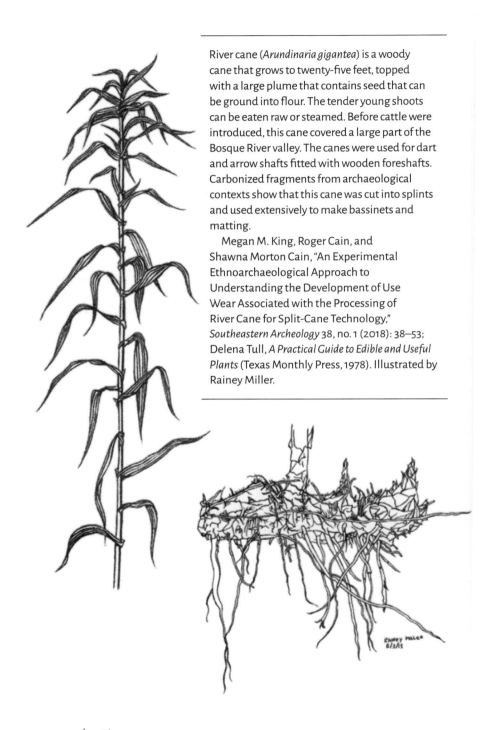

River cane (*Arundinaria gigantea*) is a woody cane that grows to twenty-five feet, topped with a large plume that contains seed that can be ground into flour. The tender young shoots can be eaten raw or steamed. Before cattle were introduced, this cane covered a large part of the Bosque River valley. The canes were used for dart and arrow shafts fitted with wooden foreshafts. Carbonized fragments from archaeological contexts show that this cane was cut into splints and used extensively to make bassinets and matting.

Megan M. King, Roger Cain, and Shawna Morton Cain, "An Experimental Ethnoarchaeological Approach to Understanding the Development of Use Wear Associated with the Processing of River Cane for Split-Cane Technology," *Southeastern Archeology* 38, no. 1 (2018): 38–53; Delena Tull, *A Practical Guide to Edible and Useful Plants* (Texas Monthly Press, 1978). Illustrated by Rainey Miller.

Rough-leaf dogwood (*Cornus drummondii*) is a small, deciduous spreading tree found in moist soil along the banks of river bottomlands from Texas to Canada. The cream-colored flowers are not very showy; the fruit is a cluster of round, white berries with a dark spot. The most common use of this shock-resistant wood was for making arrow shafts. The tree was cut or burned to ground level, from which ideal shoots would be available in a year. Medicinally, dogwood bark was used as a quinine substitute by the Confederacy during the Civil War.
  Matt Warnock Turner, *Remarkable Plants of Texas: Uncommon Accounts of Our Common Natives* (Austin: University of Texas Press, 2009). Illustrated by Rainey Miller.

goose, or turkey feathers, which were commonly split and tied to the shaft with thread-sized strips of deer sinew. The bird wing provided enough flight feathers on each wing to fletch six arrows. Each wing feather tilts in one direction from the quill, so that arrows made from the left and right wings send the missile in a perceptible drift to the right or the left; for this reason, the arrow-maker would traditionally use only the right or the left wing.[42]

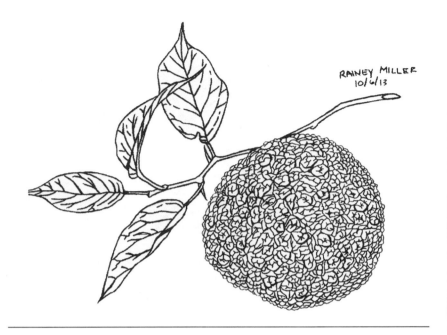

Bois d'arc, horse-apple (*Maclura pomifera*) is a hardwood tree recognized by its large, roach-repelling, grapefruit-sized fruit. The tree was widespread during the Pleistocene but nearly became extinct along with the Ice Age horses and camels that spread the seed over North America. Without dispersal agents, bois d'arc survived only in the Red River region of Texas, Oklahoma, and Arkansas. The tiny habitat was monopolized by the powerful Spiroan trade network (AD 1250–1450) that traded the bow staves all over the Great Plains. The introduction of horses resumed the dispersal of the tree. The tree became common along the Bosque River when barrels of the seed were sold for hedges for those who couldn't afford stone walls.

Connie Barlow, *The Ghosts of Evolution: Nonsensical Fruit, Missing Partners and Other Ecological Anachronisms* (New York: Basic Books, 2000); Leslie Bush, "Evidence for a Long-Distance Trade in Bois d'arc Bows in 16th Century Texas (*Maclura Pomifera, Moraceae*)," *Journal of Texas Archeology and History* 1 (2014): 51–69. Illustrated by Rainey Miller.

The bows were made from oak, hickory, mesquite, juniper, ash, and even willow, and some of these were powerful enough to have had a 60-pound draw. The most preferred wood was the hard-to-get bois d'arc (*Maclura pomifera*), which was considered the finest bow-wood throughout the Southern Plains, for a price. These bows, like the example found protruding from under a rock ledge on the upper Bosque River in the 1970s,[43] were often well-crafted and without the need for sinew or rawhide wrapping. Bois d'arc bows were made by splitting a limb and shaping the stave into a bow by scraping with chert tools, leaving knots on the back side in place. Sometimes yucca or sinew fibers were not strong enough to string the bows and twilled squirrel hide was used. The people using this new technology are associated with the new Scallorn arrowpoint style.

Recovery: The Austin Phase, AD 700–1200

After more than a century of depopulation, disruption, and reorganization, life along the Bosque River becomes more visible in the archaeological record. Floodplain analysis from Central Texas rivers from the centuries between AD 220 and AD 1050 shows that this was a rainy, stable period of increased river flow. The alluvium deposited by the river during this time is referred to as the West Range, the last floodplain build-up until the present Ford alluvium.

Increased moisture produced taller prairie grasses, which were less attractive to bison and more inviting to deer as the Western Cross Timber's canopies remained closed, providing browsing for the ideal deer habitat. The Bosque River had marshy areas and pools made by beaver dams, supporting camas meadows, waterways choked with edible cattails, and plenty of aquatic wildlife. Deer bone fishhooks found in rock shelters along the

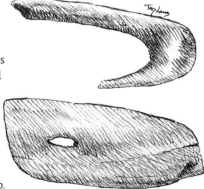

Deer bone fishhooks were manufactured throughout the Later Archaic. The leg bone was broken and shaped by carving with a retouched triangular flake of chert. Once freed from the bone stock, the hook was ground to a fine polish. This hook, associated with Scallorn points, was found in a Bosque River rock shelter by the author in 1966.

"Horn Shelter: Archaic Fishhook Manufacturing," *Texas Beyond History*, texasbeyondhistory. net/horn/manufacturing.html. Illustrated by Tad Lamb.

Bosque River during the Austin Phase suggest that fish were an important food source. Edible plants, roots, and tubers were processed in hot-rock ovens as they had been for thousands of years.[44] These were the last centuries of the Late Archaic deer and hot-rock oven lifestyle.

During this time bows became dominant as the atlatl faded from use and was generally gone by AD 800; the new Scallorn arrowpoint became the dominant type throughout most of Texas.[45] The Scallorn style was already common by AD 400 over much of the North American Midwest, Mississippi Valley, and the American Southwest, associated with early agriculture everywhere except Texas.[46] During the first part of the Austin Phase, before trade resumed around AD 1000, the Central Texas peoples who used the Scallorn arrowpoint remained in the isolated conditions imposed by the Dark Ages. While trade and migrations resumed along the Texas coast and in East Texas, Central Texas Natives remained unchanged genetically as they continued the Archaic gathering and deer hunting lifestyle.[47]

Isolation has a degrading effect on any population, and without the exhilarating benefits of trade, people living along the Bosque River must have been an impoverished lot compared to other regions. Those geographic areas lucky enough to have had an economic engine, like chert, agriculture, or bison, were in a position to bring the refreshing winds of trade and ideas into their lives. At least that's the way trade has worked throughout world history. Social mixing in Central Texas was limited to the annual gathering of the small bands coming together to share a particularly abundant food source. Regional gatherings were opportunities to stir the gene pool and for knappers to compare their particular arrowpoint and tool styles. The variety of Scallorn points most common along the Leon and Bosque River drainages, and the middle Brazos above Waco, is the fishtail-based Coryell.[48]

The Medieval Warm Period (AD 900–1300)

The Medieval Warm Period was caused by an increase in solar radiation and a decrease in cooling volcanos,[49] which brought centuries of climatic conditions favorable to the intensification of agriculture across the Northern Hemisphere. While nothing much changed along the Bosque River, these 400 years brought three brilliant cultural centers to North

America, only to have them fade when this climate regime changed: Chaco Canyon in northwest New Mexico, Paquimé in northwestern Chihuahua, and Cahokia at the junction of the Mississippi and Missouri Rivers. The Early Caddoan center at Spiro, Oklahoma, flourished, as the East Texas Caddoans spread westward, into the post oak savanna, nearly to the lower Bosque River. The Medieval Warm Period ended the cool clammy centuries in Europe and, as in North America, brought reliable crops, reinvigorated trade routes, spurred cultural awakening, and also brought more people.[50]

While much of North America was turning to agriculture and living in villages, the aquatic-based deer hunters who fished, collected mussels, and depended on wild plants along the Bosque River remained firmly entrenched in the Late Archaic lifestyle. The territorially sensitive Central Texas Natives left their Scallorn arrowpoints associated with violent death all over their range. Unwelcome encounters must have become more common as the climate began to warm and dry after 1000 and protein-hungry, growing populations outside the region began to intrude.[51] Between 1000 and 1300 large numbers of the Early Caddoans from East Texas set up intensely occupied seasonal base camps in Central Texas along the middle Brazos, Bosque, and Leon Rivers. These people were trying to meet the

---

Bow technology spread into Texas after AD 600, and the Scallorn point (AD 700–1200) was one of the earliest arrowpoints in use along the Bosque River. The Scallorn style coexisted and later replaced the atlatl-launched dart points. The point is a thin, triangular, corner-notched point with slightly concave to straight base, the examples along the Bosque River are more often slightly convex; usually less than 45 mm in length. This point was in use during a violent period of change in which the cause of death was frequently the Scallorn type.

Dan R. Davis Jr., *Prehistoric Artifacts of the Texas Indians: An Identification and Reference Guide* (Fort Sumner, NM: Pecos Publishing, 1991); Chuck Hixson, "Graham-Applegate Rancheria: What Is the Austin Phase (and the Late Prehistoric)?" *Texas Beyond History*; Ellen Sue Turner and Thomas R. Hester, *A Field Guide to Stone Artifacts of Texas Indians* (Houston: Gulf Publishing, 1999), 230. Illustrated by Jill Alford.

demands for Edwards chert and white-tail deer products among their trade contacts in the Mississippian trade network.[52]

Bonham-Alba arrowpoints, diagnostic artifacts of this East Texas intrusion, are found in Bosque River collections along the entire river, Caddoan artifacts suggesting base camp plants and deer-processing, like pottery, Gahagan knives, and bone needles, are limited to the lower Bosque and the confluence of the rivers at Waco. There is evidence of rendering fat from deer bones, collecting Edwards chert, opuntia tunas, and the most important trade item, deerskins the Caddo most valued.[53] It is not known if there was an economic partnership with the local Austin Phase people living along the Bosque River, or if the Caddo occupation involved such large numbers that the much smaller Scallorn clans just avoided them. By 1300, climate change would alter the river, and neither of these Native peoples would inhabit the Bosque River.

7

Climate Change

DROUGHT AND FLOODS

During the Austin Phase (AD 700–1200), the Bosque River was a series of wet meadows through which the river meandered, laying down alluvial sediment that had been filling the void scraped out during the droughty Altithermal centuries in ancient times. The Scallorn arrowpoint users who lived in this wetland ecosystem were the last of the Late Archaic peoples. The Bosque floodplain was covered with tall grasses and cattails, and, where the grasses had been burned away, the meadows were rich in camas, the bulbs of which could be processed and preserved for winter use. Alluvial sediment laid down during these cool, moist centuries was known as the West Range Alluvium, and it covered the Bosque River floodplain like a mantle. Beginning in the 1200s this landscape would be washed downstream.[1]

The final centuries of the Medieval Warm Period brought the alternately stormy, droughty 1200s and 1300s that destroyed the complex North American agricultural centers at Cahokia and in the Southwest, as well as the rain-dependent Scallorn-users along the Bosque River. The Medieval Warm Period delivered near-permanent La Niña–like conditions that imposed a series of decades-long droughts in 1021–1051, 1130–1170, 1240–1265, 1276–1299, and 1300–1382. The severe droughts eliminated the vegetation that had for centuries dominated Central Texas and exposed the land to the returning hurricane and tropical flooding erosional events that removed the Western Range Alluvium by the 1300s, which supported the lush Late Archaic habitat. The subsequent floodplain began filling the eroded void known as the Ford Alluvium,[2] and it is on this level surface that modern-day baseball fields have been built along the Bosque River.[3]

The Little Ice Age (AD 1300–1850)

Natural archives like tree rings, ice cores, corals, diatoms, and ocean sediment document the Little Ice Age, a very cold episode in the history of the Great Plains. This period began abruptly with strong volcanic eruptions that reduced solar intensity, altered ocean currents for centuries, and chilled the Northern Hemisphere. Grass fires coupled with droughty times led to the replacement of the rain-loving tall grasses like big bluestem with the shorter grasses preferred by bison. With the cool temperatures that mattered most to bison, the Little Ice Age was off to a chilly start with the Wolf Solar Minimum (1270–1350) completing the ideal bison habitat. By 1300 bison bones started showing up in archaeological digs in Central Texas, but the Scallorn points were gone.[4]

The Bison Return to Texas: The Toyah Period (1250–1700)

The sudden appearance of cooler weather in the late 1200s shifted the great bison herds of the Northern and Central Plains southward to the Texas Gulf Coast. The abundance of bison documented by archaeology throughout the Southern Plains shows that this was one of the largest migrations of animals in the history of North America. Moving quickly to take advantage of the new bison presence, some people migrated south from the Kansas area, and others from the Southwest, to reidentify themselves as Plains Village buffalo-hunting farmers along the Canadian and Red Rivers. The arrowpoint type used by these Indians was the Washita, a variant of the triangular side-notched point used for bison hunting all over the Great Plains. Known as the Pueblo side-notched in the Southwest, it is also identical to the Cahokian style used by Plains bison hunters who delivered meat to the Cahokian urban center before it collapsed in 1300.[5]

The Plains Villagers established an east–west trade in which they swapped their bison robes, tallow, Alibates chert, and bison wool clothing for pottery, West Coast *Olivella dama* shells, Puebloan turquoises, obsidian, jackrabbit fur, and cotton fiber.[6] On the eastern end of the trade network was the great Caddoan Spiro trade center that had survived the collapse of Cahokia, partly because this society controlled something very important—the bois d'arc bow trade. The presence of the Washita arrowpoints in Bosque River collections shows that these bison-hunting farmers took time away from their crops, probably after harvests, to hunt bison in this area.

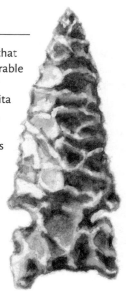

The Washita arrowpoint (AD 1200—1600) is a Plains point that spread into Texas with the bison herds drawn in by the favorable Little Ice Age conditions (AD 1300–1600). The Bosque River marks the southern border of Washita point use. The Washita arrowpoint is a small (19 mm to 27 mm in length), very thin, triangular blade with parallel notches on the lower half of the point. There were several side-notched triangular points in use in Texas during this era with subtle differences. The similar Harrell arrowpoint is longer, with a notched base.

Ellen Sue Turner and Thomas R. Hester, *A Field Guide to Stone Artifacts of Texas Indians* (Houston: Gulf Publishing, 1999), 236; William C. Foster, *Climate and Culture Change in North America, AD 900–1600* (Austin: University of Texas Press, 2012), 75. Illustrated by Jill Alford.

Bison abundance initiated a new cultural coalition in Central Texas known as the Toyah Alliance, which brought Aborigines together and combined their technologies. A unique part of the Toyah chert tool manufacturing process was the ancient use of prepared cores. This stone-working technique had not been used in most of North America since the close of the Pleistocene 12,000 years earlier. A conical core was made by striking off the upper portion of a piece of chert from which long, parallel blades were struck. Toyah camps contain small, three-inch hammerstones made from quartzite that must have been used to strike off these blades. These long, thin flakes were then used as microblades or were flaked with a deer antler into tools or arrowpoints.[7]

From the northern end of the Great Plains, bison hunters introduced their implements into the Toyah tool kit: beveled knives, end scrapers, cooking pits, and flake drills, along with the Washita arrowpoint style. Instead of continuing to use trade pottery from outside Central Texas, the Toyah potters developed their unique bone-tempered ceramics. The most popular Toyah period arrowpoint style was the small, thin Perdiz, distinguished by a contracted, pointed stem. The Perdiz has a uniquely pointed base that could have been designed to detach from the arrowshaft after impact to continue to damage tissue as the bison kept on running. Also from the upper Plains

The Perdiz (AD 1200–1500) is a small, less than two inches long, arrowpoint that is associated with bison hunting. The point is usually very thin, with sharp barbs, with a pointed stem that is a third of the length. This point's range extends from southern Oklahoma to Northern Zacatecas, Mexico. The point may have been intended to dislodge inside the running bison, to continue cutting the muscles as the animal ran.

Ellen Sue Turner and Thomas R. Hester, *A Field Guide to Stone Artifacts of Texas Indians* (Houston: Gulf Publishing, 1999); projectilepoints.net. Illustrated by Jill Alford.

was a much stronger, sinew-reinforced, recurved bow that could propel an arrow 25 percent faster than the simple bows used to hunt deer.[8]

The use of hand language, probably from Mexico, by way of the Gulf Coast, is theorized to have converged in Central Texas to facilitate the blending of local and Plains innovations.[9] The sudden appearance of the Toyah culture featuring all of these innovations could have occurred because of prestige bias: Plains hunters demonstrated the effective use of these tools and techniques, and people copied them. There was no alternative; with the world of wetlands plant-gathering gone, there was no need for manos or metates anymore. When archaeologists excavate these stones from the Toyah camps, they are burned and fire-cracked from use in cooking pits—the dietary shift was obvious.[10]

The Wichitas before the Bosque River

The first permanent agriculture-based villages on the Bosque River occurred soon after 1700, with the arrival of two groups of the Wichita peoples, the Tawakoni and the Waco at the mouth of the Bosque in present-day Waco. The Wichitas had been forced to abandon their fertile homeland on the Arkansas River in present-day Kansas and Oklahoma because of the well-armed Osage excursions from the east and Plains Apache raids from the west. These raiders collected horses and captives for their respective trade networks. The Wichitas, isolated and poorly armed, were drawn southward to the Red and then the Brazos Rivers by French traders. Operating out of Natchitoches, the French wanted the Wichita presence in Texas to serve as a buffer against Spanish encroachment.[11]

The early Wichitas peoples were Caddoan speakers who emerged from the eastern forests perhaps as long as 3,000 years ago. By the year 500, these people had divided into eastern and western groups, the Pawnees and Arikaras on the Missouri River in the east, and Wichitas and Kichais on the Arkansas River to the west.[12] When these proto-Wichitas arrived at their new home on the Arkansas River, they created a village-based hunting economy that lasted more than a thousand years. The horticultural package that they brought from the eastern woodlands—the Eastern Agricultural Complex—would ensure a stable farming and hunting lifestyle.[13]

The Eastern Agricultural Complex (EAC) centered in the middle of the Mississippi River valley, was one of the few places in the world where plants were domesticated as early as 3,000 years ago. These cultivars, before corn and beans, are known as the lost crops because nobody was growing them in historic times. One of the first of these crops was wild rice (*Zizania aquatica*), which had been cared for around ponds and slow-moving water for millennia without morphological change—a sure sign of domestication. When rice was grown along marshy creeks, the nascent farmer would have found that weeding and transplanting seedlings to new, cleared areas would increase productivity. The addition of other useful grains to this riverine system was the logical next step.[14]

An early domesticated grain in the EAC was sumpweed (*Iva annua*), a stinky plant that still prefers streams, ponds, and sloughs, with (32 percent protein) seeds four times larger than the extinct ancestor (*Iva annua macrocarpa*).[15] Lamb's quarters (*Chenopodium berlandieri*) and its nearly domesticated subspecies (*Chenopodium berlandieri* ssp. *jonesianum*) might have been the most popular food plant documented from cooking residues, storage pits, floors of houses, and human paleofeces. Related to the currently popular quinoa, this plant grows to six feet and can be harvested as a salad green or cooked like spinach. Early explorers reported that lamb's quarters was harvested in late summer for the tiny seeds that were ground into flour and baked into bread. Little barley (*Hordeum pusillum*) was collected for a millennium without showing any sign of domestication. Two other starchy-seeded plants that were domesticated enough to die out when corn and beans were introduced were maygrass (*Phalaris caroliniana*) and erect knotweed (*Polygonum erectum* ssp. *watsoniae*). Of these ancient cultivars, only the acorn squash/pumpkin/gourd groups (*Cucurbita pepo* ssp. *ovifera*), and sunflower (*Helianthus anuus*) survived the chaos and demographic collapse caused by European contact.[16]

The Wichita Plains Villagers began to grow a poorly adapted corn variety as early as 500, but only as a minor crop. The first patches of corn traded up from Mesoamerica by way of the Southwest could have been grown for years without freshening the gene pool from additional trade and suffered from "inbreeding depression" resulting in lower productivity. It wasn't until the favorable conditions of the Medieval Warm Period (900–1300) that better-adapted corn varieties became an important crop.[17] Corn, however, did not contain all the nine essential amino acids needed to deliver a complete protein, making it fully nutritious. It needed to be eaten with a lysine-rich food like beans or rice to provide the essential amino acids needed to combine with corn to make a complete protein. Before beans were introduced centuries later, wild rice filled this role.[18]

The common bean (*Phaseolus vulgaris*) had been around in the American Southwest for centuries before finding its way to the Northern Great Plains, to the Great Lakes, and then down to the Deep South and then west to the Caddoan people of Oklahoma and East Texas. This trek took nearly four hundred years to reach the Wichita villages right next door to the Puebloan source. An interesting reason for this loop must have been cultural barriers that kept beans confined to the Southwest until the 1300s, too late for Cahokia or the Mississippi River valley chiefdoms that had collapsed by this time. This centuries-long trade barrier must have been the long Athapaskan (early Plains Apache) migration that began streaming southward out of Canada in the 800s, until they occupied the Southwest and Southwestern Plains by 1300.[19]

These early Wichitas, along with the rest of the Caddoan world, survived the AD 1300s dissolution of the iconic Chaco Canyon and the Cahokian chiefdoms. The elite-centered nature of their trade suggests that both of those societies were tempted by the Mesoamerican model of a self-absorbed elite that was blind to the needs of their people. The Chacoan people escaped the corrupt, theocratic authority of the Great Houses by

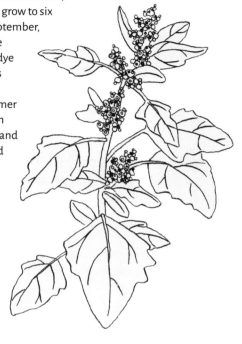

Lamb's quarters, Indian spinach (*Chenopodium album, C. berlandieri*) is an annual herb that can grow to six feet, with alternate leaves, and after September, flowers form as clusters at the top of the plant. Some Natives used the leaves to dye their arrows green. Lamb's quarters was an ancient food plant that was double-harvested; young spring and early summer leaves were cooked like spinach or eaten raw, in the fall the seeds were gathered and ground into flour to be baked into bread in the winter. Paleobotanists believe that a *Chenopodium* cultivar was traded northward from Mexico and has since faded into the general species. The seeds are equal to corn in calories, and the cooked leaves have vitamin A and C and three times as much calcium as spinach.

Kelly Kindscher, *An Ethnobotanical Guide: Edible Wild Plants of the Prairie* (Lawrence: University Press of Kansas, 1987). Illustrated by Jill Alford.

taking advantage of the late AD 1200s droughts to rebel and move away.[20] Closer to home, the rise of warlike feudal chiefdoms along the Mississippi River valley after Cahokian influence faded could have led to violence and conquest in the Caddoan world. What saved the Plains Wichitas, Spiro, and the rest of the unfortified Caddoan societies was their egalitarian government structure. There was no hated elite to rebel against. Granaries in Chaco and Cahokia were centralized and controlled by the autocratic leaders, while in the nonurban Caddo/Wichita settlements corn was stored in individual farmsteads, as was political power. The caddis and other community leaders were elected among the Caddos, and the mounds were for public buildings, not important people as with the Cahokians along the Mississippi.[21]

The isolated, declining, and contentious Mississippian chiefdoms were well-fortified with log palisades and ditches, but in the Caddoan confederacies there were no fortifications around the villages. Caddoan population areas could be visualized as "concentric rings," with the outer zone recognized as hunting territory and farming villages with their caddis, and they surrounded a ceremonial center and home of the priesthood. There were border authorities who watched the access roads to the center, triggering a sophisticated military response if needed. The Mississippian societies continued to squabble and retribalize while the Wichitas continued on their path to Texas.[22]

Psychoactive Herbs

The Toyah period in Texas, like the rest of North America after the collapse of Chaco and Cahokia, was a time of constant encounters among ethnicities. Diplomatic rituals involving the smoking of tobacco were developed to facilitate these interactions. Wild tobacco (*Nicotiana* sp.) was available in North America for millennia, but there is no evidence that this harsh variety was used. A less caustic South American cultivar (*N. rustica*) was among the large-leaf, high-nicotine cultivars introduced into North America along with corn. Before tobacco, and in addition to it later, were perhaps a hundred plants smoked in a "staggering pharmacopeia" of psychoactive herbs. Though smoking was a North American invention, the shamanistic predisposition toward psychoactive plants was brought over from Asia, where it was no doubt already ancient. At some point, perhaps among the Athapaskans in the Northern Great Plains, a pipe ritual developed for purposes of mediation and alliance. Tobacco was most likely chosen because of its properties of anxiety reduction and heightened mental acuity experienced by nonaddicted, infrequent users. The "Puebloan-Southern Plains-Wichita-Caddoan macroeconomy" was maintained by the calumet ritual despite nearly constant violence until the Spanish arrived in the sixteenth century.[23]

Before tobacco there was an herb used to facilitate supernatural diplomacy with gods rather than men; the herb used for these encounters was datura (*Datura stramonium*). Several species of sacred datura grow all over North and South America, and its hallucinogenic use must be very ancient. The earliest documented use, 5,000 years ago, is shown by carbonized seeds found in a rockshelter along the lower Pecos River.[24] But the use of datura as a divine beverage is thought to be much older. Oral histories collected in the early twentieth century record a common use in most American

Sacred datura, locally known as jimsonweed, native to the Bosque River and the southern United States and northern South America. Perhaps the most dangerous of the psychotropic plants used by Aboriginal Americans to invoke supernatural helpers through a hallucinogenic experience.

    Matt Warnock Turner, *Remarkable Plants of Texas: Uncommon Accounts of Our Common Natives* (Austin: University of Texas Press, 2009); A. H. Gayton, "The Narcotic Plant Datura in Aboriginal American Culture," PhD dissertation, University of California, Berkeley, 1926; and Weston La Barre, *The Peyote Cult* (Norman: University of Oklahoma Press, 1938). Illustrated by Rainey Miller.

---

cultures in vision-seeking and communication with the gods.[25] It's conceivable that datura-induced hallucinations were used to experience the sacred narratives of creation and perhaps explain why the Wichitas referred to the gods as "dreams."[26] The story of creation is very dreamlike: the Great-Father-Above started with land floating on water, no light, no stars, and no people until he formed the earth and sky. The First People were given the power to dream what they needed—and then they would have it.[27]

Climate Change | 81

Later, a slightly less dangerous psychedelic, the mescal bean (*Sophora secundiflora*) commonly known as the Texas mountain laurel, became the preferred agent for divine communication. There is general agreement among ethnohistorians that even though the tenets of the Wichitas' religion are rooted in southeastern tradition, it was the Wichitas that began the use of the unnaturally bright red mescal bean to provide that divine spark to ceremonies. Since the red bean grows only in Central, South, and Southwest Texas where it has a long history, the Wichitas must have found out about it from the south. From the Wichitas, the ritual use spread throughout the Plains. The shiny red bean was so valued on the Plains that as late as 1820, ten beans could purchase a horse.[28]

Yaupon (*Ilex vomitoria*). Today this shrub is one of the most common landscape plants in Texas with a long history of use by Native Americans. Yaupon tea is the only caffeine-based drink native to the state. The stimulation of the drink made it an ideal part of the Calumet ceremony, used as a negotiating agent. European observers were more impressed with the use of this drink when it was brewed strong enough to become the *black drink* used as an emetic to ritually purge the body.

Matt Warnock Turner, *Remarkable Plants of Texas: Uncommon Accounts of Our Common Natives* (Austin: University of Texas Press, 2009); Charles M. Hudson, ed., *The Black Drink: A Native American Tea* (Athens: University of Georgia Press, 1979). Illustrated by Rainey Miller.

In the thirteenth century, the La Junta–based Jumanos opened a southern trade route to avoid the Plains Apaches and connected with the Caddoan world. One of the first trade items was the mescal bean, which became part of the most important Mississippian-inspired ceremonies. The beans had to be "killed" by baking them on a stone near the fire until they turned yellow; then the beans were ground into a powder that was brewed into a sacred tea. After the people danced until midnight, the tea was consumed as part of the Feast of the First Fruits Ceremony.[29]

As was common in many Mississippian and Caddoan cultures, there were ceremonies of renewal among the Wichita peoples. The most noted of these was the Feast of the First Fruits. In late spring, when crops were nearly ready to start producing, no one was allowed to pick or eat anything until a priest was notified and the ceremony was planned. Anyone breaking this rule was almost certain to be bitten by a snake soon after.[30] The village streets would be swept and the women would clean the council house, preparing sitting mats and covering the floor with wildflowers. The ceremony began with all of the village members of both sexes gathered in the large grass dome, a sixty-foot-wide council house. Central to the celebration was yaupon tea (*Ilex vomitoria*), a caffeine-based drink normally associated with the Calumet ceremony or socializing generally, but for this ritual, it was brewed so strong that the color was black, instead of yellow, and it became an emetic. And like coffee, the caffeine absorption rate of the *black drink* was thirty times higher when the brew was served very hot.[31]

Deliberately induced vomiting was practiced all over North and South America, with the same purpose: ritual purification to get rid of past contamination.[32] Each spring was a time to start over, and the first cultivated crops were the new foods for a renewed life. But the new foods couldn't be eaten until the accumulated pollutions of guilt and misdeeds of the past year were expelled through the purge provided by the black drink. After the community purge, the priest relit his pipe with new tobacco and walked among and embraced the people as the feasting began.[33]

8

Disease Changed Everything

When Francisco Vazquez de Coronado bullied his way into the Wichita villages in 1541, the event marked the beginning of the end of the Wichitas. Following on Coronado's heels, wave after wave of epidemics swept over the people. By 1901 the population had dropped from 200,000 to less than one thousand. European diseases were raging in Mesoamerica. After the initial smallpox epidemic launched by the conquest in 1520 came the even more lethal epidemics of *cocoliztli*, from *Salmonella enterica* subsp. *enterica serovar Paratyphi C*, in 1545 and again in 1576. This disease killed as many as 90 percent of the Mixtec population in Mexico—more than twenty million people.[1] The hemorrhagic *cocoliztli* spread northward in the 1620s, killing as many as sixty thousand Puebloan people along the upper Rio Grande, and then east to the Apaches; it's not known if the Wichita were involved in this contagion. Waves of influenza, mumps, common colds, typhus, typhoid, measles, cholera, and especially smallpox streamed northward into the Plains over the next centuries.

The ancient trade routes that occasionally made the Wichitas prosperous brought European diseases into the interior and began to decimate the Indigenous population even before the arrival of the Spanish to the region. In following decades before renewed European contact, "unknown diseases" thought to have been measles and smallpox wreaked havoc on the Texas and the Caddoan peoples.[2]

French Trading Partners

In the eighteenth century, when horses and guns empowered their enemies, the Wichitas were driven to escalate the bison robe trade with the French in exchange for arms.[3] By the early 1700s the Wichitas were dependent on French flintlocks and powder, always in short supply. This trade unfortunately exposed them to more frequent epidemics of Old World diseases. During the Maunder Minimum, the coldest part of the Little Ice Age in 1706, a smallpox epidemic collapsed the already broken Jumano trade route between the southern Pueblos and the Wichitas. The ancient networks were soon replaced by new pathways that served the Spanish and French trade. For a time in the mid-eighteenth century, the Wichitas became the principal actors in the Southern Plains trade.[4] In 1747 the French were able to open trade with the Comanches by negotiating an alliance with the Wichitas' middlemen. The alliance with the Comanches covered the Wichitas' west flank as they moved farther south to the prime trade centers on the Red River.[5]

Prosperity might not be the right word for the accelerated eighteenth-century commercialization of the Wichita's lifestyle. To meet the worldwide demand for bison hides, the matrilineal family unit became a cottage industry that served the market-minded French. Intensifying trade was necessary to acquire more firearms to protect the people. The Wichita men abandoned tradition and sought additional wives to increase household production, thereby reducing the women's social status.[6] Cultural traditions were simplified as leaders and shamans appeared helpless in the face of constant trauma. The calumet pipe tobacco ceremony was used for conflict resolution among subgroups and to ease the culture shock of assimilation into fortified, pluralistic villages. Sometimes they resorted to the use of a new peyote ritual (*Lophophora williamsii*) to invoke the sacred to ease aggregation anxiety and rethink their cultural imperatives, in ceremonies moderated by a hallucinated female deity.[7]

After a 90 percent drop in population by 1700, the remnants of the Wichita bands reinvented themselves through consolidation and a shared language identity. There were a new people, mounted, thanks to the Pueblo Revolt of 1680 that spread horses throughout the Plains, and began to acquire firearms. The Wichitas were now frequently able to tip the balance of power their way. Beneficial calumet-facilitated alliances opened Wichita's access to horses that were delivered to the French along with buffalo products in return for more muskets and manufactured goods. However, unlike the well-supplied Osages, the Wichita bands were so far inland that the resupply of powder and lead was too random for the serious business of defense.[8]

The Wichitas Arrive in Texas

Several Wichita bands joined the Tawakonis and moved south to the Red River in the 1720s, to be in contact with French traders by water as middlemen in the east–west trade traffic. Soon the aggressive Osages extended their range to raid the Red River settlements, forcing Wichita bands to move again. Eventually, the Tawakonis and Wacos built two villages at the present location of Waco, Texas, at the junction of the Bosque and Brazos Rivers. The Waco's village was built on the left bank of the Brazos River, just above the "Waco Spring," where cold water rushed out of cracks created by the Balcones Fault.[9] When the iconic grass dome houses were built, around four acres were planted with corn, beans, squash, and watermelons. Ancestral crops were no longer reported, not even the sunflower. The peas mentioned were almost certainly the Chickasaw pea described in Southern agricultural publications as a red variety producing a bushel of dried peas per half-acre in cultivation. As an early (*Vigna unguiculate*) cultivar, this black-eyed pea ancestor (really a bean) is a rampant climber, producing nutritious horse fodder.[10]

Observers mentioned peaches planted by Wichita horticulturalists at Waco; these must have been the Indian Blood cling and similar types introduced through the Spanish missions. Easily transported and planted, the seeds had been traded all over eastern North America by this time.[11] These peaches thrived, even going feral a century before imported diseases destroyed their vigor and moved them from cultivation to the occasional pink bloom seen in chaotic thickets. These peaches, like the other crops, were left in the charge of those unable to make the early summer trek to capture horses and hunt bison.[12]

There was a problem with these western excursions for bison and horses: the Lipan Apaches held the territory from west-central Oklahoma south to Mexico. For centuries the Apaches had raided the Wichitas, for captives and horses, blocking access to the west generally, using mounted fighting tactics learned from the Wichitas themselves.[13] Concerned for their western security in Texas, the Wichitas began to invade the Lipan homeland on the Colorado River as early as 1740. The Lipans identified the raiders to the Spanish in San Antonio as some sort of Caddoan group. Outgunned and outnumbered when the Comanches allied with the Wichitas, the Lipans became defensive and looked to the Spanish for protection.[14] Not realizing the size of the powerful Comanche presence on the Southern Plains, the Spanish constructed a presidio and mission on the San Saba River to Mexicanize the Lipans and establish a gateway to the Plains. An attack on the mission in 1758 by an alliance of two thousand combined (Taovayas) Wichitas, Comanches, and others, destroyed the San Saba mission and ended Spanish plans for northern expansion.[15]

Rather than aggression, the Spanish response was to double down on their efforts to keep French trade, especially firearms, from reaching the Wichita groups. The Spanish policy of maintaining the status quo was at odds with the profit-driven French traders who gladly traded firearms for hides and horses. The Spanish officials were relieved, at least at first, when at the end of the French and Indian War in 1763, French Louisiana passed to Spanish control.[16] Among the French citizens invited into the service of Spain in 1769 was Athanase de Mézières, the fur trader turned diplomat who was popular with the now-impoverished Wichitas and Caddos. Mézières was able to stabilize the Texas frontier for decades by using the French policy of fair dealing, plenty of presents, and a generous flow of goods. The Wichita elders used the gifts to reward good behavior among the young men; when the gifts didn't arrive, the youth sought honors through raiding.[17]

Even with the French gifts-and-trade model, Spanish goods were not sufficient to meet the needs of the Wichita people. The Spanish reluctance to trade firearms placed the Wichitas still above the Red River in a dangerous situation. Illegal French traders in the Arkansas River valley traded English muskets to the Osage raiders, who pushed farther south each year to raid and steal horses to buy more firearms.[18] These Wichita bands responded by scattering southward from the Red River villages; some of them moved to the Bosque-Brazos River confluence to add their roundhouses to the Tawakoni and Waco villages.[19] There were six villages mostly in what is now the downtown area of Waco. The village of El Quiscat, named after the noted Tawakoni mediator, was located a few miles north.[20] In the 1770s Mézières held peacekeeping meetings of erstwhile norteños in El Quiscat, where promises for Wichita coordination against hostile peoples and increased trade were made and later blocked by Spanish bureaucratic inertia and suspicion.[21]

European Diseases Change History

The history of Spanish Texas took a catastrophic turn in 1777 when an especially virulent smallpox (*Variola major*) swept through for the first time in decades. Though deadly to some Euro-Americans (Mézières lost his wife and two children in a week), the virus was more fatal to Indigenous peoples with little exposure or acquired immunity to European diseases. The smallpox virus is very adaptable to most immune systems, and, in people with similar immune systems, the pathogen becomes preadapted to burn through a population. To the virus, most Indigenous Americans were like siblings. Many Indians died within days, even before the typical body sores could develop. Less affected were the American colonists and British in the Siege of Boston (considered the point of origin in 1775), during the first year of the American Revolution.[22] George Washington was able to protect his army and control the virus by quarantine, care, and inoculation, aided by centuries of acquired immunity.[23] None of these procedures were known or available to the Wichita bands; they lost a third of their population.[24]

The epidemic was an opening for Osage raiders to increase pressure on the Caddoan-Wichita peoples who received neither firearms nor military assistance from a weakening Spanish government.[25] In the early 1780s increasing Anglo-American activity east of the Mississippi River began to crowd out the Choctaw people and drive them west into Louisiana and East Texas. The Spanish governor of Louisiana received these well-armed Indians and asked the Caddos to share their hunting grounds, noting that the Choctaw were bitter enemies with the Osages and would provide a measure of security. In a few years the Choctaw incursion became an effort to muscle into East Texas to settle permanently among the Caddos.[26]

Practically unarmed in the face of these hostilities, the Caddos, Wichitas, and affiliated peoples found a way around the Spanish firearms boycott by inviting in their eventual destruction. At the end of the American Revolution in 1783, adventuresome Anglo-Americans turned their attention to establishing trade with these desperate Texas Natives, ignoring Spanish authorities in Louisiana and Texas. Most notable of these was Philip Nolan, who supplied firearms to Waco-based and other Wichita-related groups from 1791 to 1800, in exchange for assistance in trapping feral mustangs farther up the Bosque, Brazos, and Colorado Rivers. Wichitas, Caddos, and the Comanches formed an alliance to block further Choctaw incursions. During these years the Choctaws attacked the Louisiana Caddos with war parties numbering in the hundreds, forcing these Indians west to the Trinity River in Texas. When Nolan was killed by a Spanish patrol, he had with him Anglo and Cherokee wranglers who spread the word about Texas opportunities.[27]

Anglo-Americans Discover Texas

Napoleon's European conquest forced the Louisiana Territory back into French control, which was then purchased by the United States in 1803. As Spanish relations with Caddo, Wichita, and allied groups weakened, American trading companies set up operations. A major downside of trade with the mostly Scots Irish Americans was alcohol. Illegal liquor traffickers stayed ahead of licensed traders, collecting hides in exchange for whiskey.[28] Whiskey was the preferred contraband because it would be consumed quickly, creating a demand for more.

After trading alcohol to Native peoples east of the Mississippi River for generations, whiskey peddlers knew that alcohol served two functions: valuable furs and other commodities could be obtained cheaply, and the debilitating effects of alcohol worked to ethnically cleanse new areas. Some Indigenous Americans, like their Northeast Asian forebears, carried a genetic vulnerability to alcohol in the same way that they lacked protection from other Old World diseases.[29] The biological predisposition was compounded by the trauma of constant raids and epidemics among Native children, which drove their alcoholism toward antisocial, destructive behavior.[30] The French and Spanish considered their relations with the Indians as long-term and avoided including alcohol in their trading relations. This was not the case with American trader's predatory capitalism.[31]

The successful conclusion of the War of 1812, followed by the Panic of 1819, released a flood of Anglo-Americans moving westward hoping to renew their fortunes. Moses Austin arrived in San Antonio in 1820 with a unique proposal to bring in Catholic families from Louisiana to form a buffer zone to block the flood of less favorable adventurers. Austin died after some success in the approval process, and his son, Stephen F. Austin, created a sensation as cheap Texas land was advertised. Settlers arrived at the mouth of the Brazos and Colorado Rivers clamoring to populate his new colony. Austin had spent a year in Mexico City negotiating his way through the turbulent politics to get his colonization project approved.[32]

The Texas Natives living at the center of Austin's colony were the Karankawas, considered peaceful until the arrival of Anglo-Americans in their midst. Naturally, there was some conflict, leading Austin to declare in 1825 that settlers should "pursue and kill all those Indians wherever they are found."[33] In these first conflicts with the Indigenous Texans, the concept of the Texas Rangers began as the enforcers of ethnic cleansing.[34] Driven by fabrications of cannibalism and murder, a large force of colonists attacked a Karankawa camp at the mouth of the Colorado River.[35] The poorly armed Karankawas tried to swim across the river to escape as men, women, and children were killed by gunfire.[36] The ease with which Karankawas were exterminated set the lasting genocidal pattern of Scots Irish and Anglo-American contact with Texas Indians. "Indian hunting had become a sport in Texas," which included "chasing squaws" who became temporary sex slaves.[37]

The Anglo-Americans were most favored by Austin because they paid their fees and established their plantations in the river bottoms. These Anglos had arrived in Virginia in 1655 as aristocratic refugees in the decade after they lost the English Civil War. Like the other British migrations, they set about reconstructing the nearly medieval way of life they had known as Royalists in the south of England—known in Britain as Cavaliers. This small elite soon dominated Virginia, and their slave-owning, peasants-be-damned mindset, developed over centuries, shaped the social history of the Deep South and East Texas. Profoundly conservative, "they detested every nation except England and despised every race except their own."[38] These Anglos were ideal agents for Austin's plans for Texas.

The Scots Irish Arrive in Texas

Among others arriving in Texas were the Scots Irish, mostly from the upper South and too poor to pay Austin's land fees. Their attitudes and values were destined to define North Central Texas well into the twenty-first century. These people were from the British borderlands of northern Britain, including the ancestors of English occupiers of Northern Ireland and those along the borders of Scotland and Britain. Irked by politics in their homeland, these people became a great folk migration of 250,000 immigrants first on the East Coast and then quickly into the frontier zone during the eighteenth century. The Scots Irish were considered scandalous in their felt hats, belted trousers of the men, and even more so in the women who presented full bodices and bare legs visible under short skirts.[39]

Referred to as "the scum of the universe," the Scots Irish settlers headed for the far frontier where land was free. During their time in the backcountry, they shed their fiery Calvinist Presbyterian faith for a fiery evangelical, mostly Baptist and Methodist religious identity. Frontier religion was not a moral agency; sermons were filled with military metaphors and prayers for vengeance and the destruction of enemies. Armed with a Hebraic vision of acquiring promised land through conquest and continual forgiveness for atrocities, the Scots Irish projected violence onto the frontier. Their character was forged by seven hundred years on the British borderland where Scottish and English kings fought back and forth, creating a culture of violence where preying on their neighbors became a way of life. The culture of the Scots Irish was shaped by anxiety, xenophobia, intense pride, an obsession with weaponry, and valued force over reason.[40]

When the Pennsylvanian forges began to produce a lighter version of the central European rifle, the Scots Irish saw it as the must-have weapon. Soon called the Kentucky rifle, this weapon could strike a target at two hundred meters, far superior to the smooth-bore musket lethal only to about sixty meters. The Scots Irish invaded Kentucky in the late eighteenth century, regarding Native resistance as justification for further violence. They built the log cabin introduced by the Scandinavians, responded to Indian resistance with casual genocide, and before long moved farther west and repeated the process until they entered Texas.[41] This westward movement and disregard of Natives as real people has been called Manifest Destiny by historians; it was, and still is, a sense of divine purpose that favors the Scots Irish over all others.

Austin's plans for ethnic cleansing, even with the loss of land fees, meshed perfectly with the Scots Irish, who couldn't have been stopped anyway. The Karankawas, already weakened, were brushed aside easily, but the Wichita villages at the mouth of the Bosque River were another matter. Austin realized that these villages blocked his plans for settling land farther up the Brazos River, but he feared their military prowess, even if they were almost completely without firearms. The answer to his problem was typically ruthless. Cherokees had been migrating into East Texas since before Austin's Mexican contract, and they had been in Mexico City while he was there, trying to get official recognition for their land. But US Minister Joel Poinsett, who despised Indians and Mexicans equally, had the ear of those in power, and the Cherokee petition was ignored. Once back in Texas, the Cherokees were trying to get some sort of recognition from Austin for their land. Austin, thinking the enemy of my enemy is my friend—at least until I kill him—asked the Cherokees to prove to their new friends how useful they could be in return for legal possession of their land.[42]

No Safety in Treaties

The Tawakonis and Wacos signed a treaty with Mexican authorities in 1827 and felt secure in their villages, but the North Texas Comanches and Taovayas had not signed the treaty and continued to raid Austin's colonies. Mexican officials and a dozen Anglos entered one of the Waco villages while all but a few elderly people were away for the bison hunt. The Mexicans discovered that these Indians were not the hostile parties but had to restrain the Anglos who were offended at not being allowed to kill these Wacos. A few weeks later two hundred of Austin's Rangers located the Waco hunting camp on the San Saba River and immediately charged. Waco scouts had seen them coming and evacuated almost everyone; the Rangers killed one old man and captured a few women and children.[43] The Wichitas retaliated by stealing horses from Austin's and other colonies, providing the justification to call in the Cherokees.

In 1829 a sizable force of Cherokees crept up to a Waco village near the mouth of the Bosque River. After generations of contact with Anglos, the well-dressed Cherokees, armed with long rifles, regarded the Wichitas as savages. They had no idea that they would become the target for extermination in just nine years. As they closed in at dawn a barking dog alerted the village, and then the Wacos, armed mostly with bows and arrows, retreated with heavy losses. The Cherokees killed women and children as well as men to destroy these people. The last of the Wacos gathered for a last stand when a couple of hundred mounted Tawakonis showed up but were afraid to get too close to the Cherokees' rifles. Thoroughly outnumbered, the

Cherokees retreated.⁴⁴ The Cherokees returned in 1830 to attack the nearby Tawakoni village, driving them with accurate rifle fire into their bunker. The safe house was a ring of stones a few feet high with bison skins on a pole frame; the Cherokees used burning bundles of grass to drive them out to be massacred.⁴⁵

Waco and Tawakoni survivors retreated west to start another village, but in 1831 this village was surrounded by the Mexican army and destroyed, along with fields of corn and beans. Among the dozens killed were Waco leaders who had urged peaceful relations with the Mexicans and Anglos. Devastated, these Natives kept moving westward, reestablishing camps and fields that were destroyed before crops could be harvested. Some of the young men joined the Comanches in revenge raids, keeping the cycle moving.⁴⁶ Hounded by Texas Rangers and Scots Irish raiders, most agricultural Indians ended up in North Texas, where they were met with the supreme act of ethnic cleansing.

In September of 1835, as Texas Rangers had just burned another Waco village on the Trinity River in the Dallas area, a man named James Neil injected the smallpox virus into a single Waco captive who was then released and allowed to join the fleeing villagers. "Neil was never able to ascertain the success or failure of his little experiment."⁴⁷ Maybe James Neil never heard that his little experiment ignited a western North American pandemic that killed Indians from New Mexico to Canada. Just as Americans were beginning to survive smallpox epidemics (*Variola major*), Neil's little vial contained black smallpox (*Variola hemorrhagica*), also known as *rotting face*. Hemorrhagic smallpox can darken the skin, burst the capillaries, and cause bleeding from the mouth and internal organs.⁴⁸

It was probably the Comanches who sheltered the Wichita refugees and then carried the disease northward, killing unknown thousands in the smallpox epidemic of 1837–38. Ninety-eight percent of the Mandans died, twenty thousand on the upper Missouri, along with half of the Assiniboines, and many Plains Crees. So many died that the bodies could not be buried.⁴⁹ As the Wichitas retreated away from Texas, they must have been mindful of their religious prophecy that described the last level of their existence, to wait for a great star to point to the man who would restart the four-cycle stage and a new, vibrant beginning.⁵⁰

9

The Comanches and Euro-Americans

Once again it was the climate that set into motion shifting habitats, migrations of animals, and dislocations of peoples that brought a new culture to the Bosque River. This was the iconic Comanche, mostly remembered for their cameo roles as the savages in cowboy movies. The Comanches raided along the Bosque River in the mid-nineteenth century and traded brutalities for brutalities with the better-armed, mostly Scots Irish immigrants. They were the last vengeful survivors of a once great empire that had stretched from the central Great Plains to San Antonio.[1] In 1770 Comanches numbered around forty thousand, dropping to less than ten thousand in the 1850s, then down to fifteen hundred in 1875 following military defeats, diseases, and loss of the bison amid a long drought.[2]

It was during a drought that archaeologists started to pick up the artifact trail of the Shoshone mountain people, the parent group of the Comanches. Toward the Medieval Warm Period (900–1300), increased solar activity heated the earth and nearly depopulated the western Great Plains. Things were less dire in the Great Basin and Rocky Mountains where the Shoshones divided their year between fishing, hunting deer and mountain sheep, and gathering food in the summers. In winter they would sometimes use the South Pass (Wyoming) to access the eastern side of the Rocky Mountains to find bison and elk sheltering from a series of intense droughts in the thirteenth century. These decadal droughts cleared most people from the Plains, creating a void that the most adaptable Indigenous peoples would be pulled into when the climate began to favor bison.[3]

Puccoon (*Lithospermum incisum*) is a perennial herb growing to a little over a foot tall on an erect stem. The bright yellow, trumpet-like, fringed flowers bloom from early spring to June. The preferred habitats are dry, sandy prairies and open woods. Texas Indians used the roots for tea to treat stomach trouble and diarrhea. Several species were used as contraceptives for women; when a course of treatment was followed, sterility became permanent. A species of puccoon that contains a natural estrogen is currently being studied as a simple and inexpensive method of birth control. Natives burned the top of the plant as ceremonial incense and the root provided a violet to red dye.

Kelly Kindscher, *Medicinal Wild Plants of the Prairie: An Ethnobotanical Guide* (Lawrence: University Press of Kansas, 1992). Illustrated by Jill Alford.

The northern Great Basin, home of the Shoshones, filled up with snow as the Little Ice Age (1300–1850) grew in intensity around 1450 and drove the people and the bison into the Great Plains. The drought and attendant grass fires had cleared the trees and brush, which helped spread little bluestem (*Schizachyrium scoparium*) to the Texas Gulf Coast.[4] The short grass habitat of the Great Plains recovered from the dry years through a hierarchal succession of the most drought-tolerant grasses that formed communities nursing the next, more valuable occupant. The first to cover bare ground was buffalograss (*Bouteloua dactyloides*), blue grama (*Bouteloua gracilis*), and Texas grama (*Bouteloua rigidiseta*) This stage was followed by the more nutritious side oats grama (*Bouteloua curtipendula*) that soon crowded out the first-responders until the succession ended with the final establishment of little bluestem. The Shoshones reinvented themselves as Plains hunters as they followed the bison south.[5]

Shoshone/Comanche arrowpoints are triangular shaped like the standard Pueblo to Plains side-notched/unnotched type. The Shoshone/Comanche point is a trinotched style with flaring barbs on the sides of a concave base. This point, which resembles so many Texas styles, was dropped from use before these Indians reached Texas when the metal trade point became common. And then there is the Intermountain Ware, a ceramic style resembling a flowerpot, that also traces the path of the Shoshones until the Comanche split away from them.[6]

Ethnographers have collected stories that refer to the break-up that mention horses, smallpox, and an argument. The horses arrived among the Shoshones from the untold thousands of horses made available from the Pueblo Revolt of 1680 and from the additional 10,000 captured by Apaches from the Spanish refugee camps around El Paso in the next five years.[7] Following the horse traders northward from the Spanish Southwest they encountered a double epidemic of measles and smallpox in 1692–93, which resulted in stress and contention for reorganization.[8] Shoshone legend contains several violent incidents that sparked the animosity that led the Comanches to split away from the Shoshone. After the split, the Comanches moved south for better access to more horses and bison.[9]

The Comanches, or the Numunu, followed the front range of the Rocky Mountains into the Southern Plains. They had learned mounted fighting and specialized in equestrian bison hunting as they migrated. As newcomers to the region, they had ridden into the last days of the busy, centuries-old trade networks between Pueblos, Apaches, Wichitas, and Caddos, facilitated by Jumanos, just as the network was weakening. The Comanches' southwestern advance was noted at Taos in 1706, although their middle movement into Texas was not recorded;[10] it was probably not long after. The Apaches controlled the Southern Plains when the Comanches arrived. The Lipan Apaches had recently moved into the Southern Plains to raid Wichita villages for slaves for the Spanish market in New Mexico. In the early 1700s, the Lipans were overwhelmed by the Comanches and escaped south into Central Texas where they found temporary safety,[11] perhaps along the Bosque River. Then the Comanches reinvented themselves as Lords of the Southern Plains and organized their commercial networks.[12]

Feral Horses and Cattle along the Bosque River

Horses were as important as bison, and there were plenty of feral mustangs descended from desert-bred African barb horses, as well as tempting Spanish stock. There is a story, probably apocryphal, that in 1690 Alonso de León left a "bull and a cow, a stallion and a mare at each river crossed" as he traveled from the Rio Grande to establish his mission in East Texas. By 1716

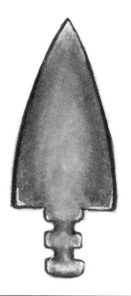

As the Shoshone peoples were migrating southward in the late 1600s, they were using the Shoshone trinotched arrowpoint—a chert point that was similar to other Western triangular styles of the period. By the time the group of Shoshones who had horses became known as the Comanches and arrived in the south, they were using metal arrowpoints. Some of these metal points were chiseled from barrel hoops and other sheet metal, and some, like the one illustrated, were trade points.

Cody Newton, "Towards a Context for Late Precontact Culture Change: Comanche Movement Prior to Eighteenth Century Spanish Documentation," *Plains Anthropologist* 56, no. 217 (2011): 5; William E. Moore "A Guide to Texas Arrowpoints," *Brazos Valley Research Associates, Contributions in Archeology*, no. 6 (2015): 51. Illustrated by Jill Alford. "Late Precontact Culture Change: Comanche Movement Prior to Eighteenth Century Spanish Documentation," *Plains Anthropologist* 56, no. 217 (2011): 5; William E. Moore "A Guide to Texas Arrowpoints," *Brazos Valley Research Associates, Contributions in Archeology*, no. 6 (2015): 51. Illustrated by Jill Alford.

Spaniards reported black Castilian cattle and horses along the Trinity River and likely throughout Central Texas. The Apaches and Comanches hunted the cattle, called *cimarrones*, which were said to have been more difficult and dangerous to hunt than bison or deer.[13] These cattle browsed along the Bosque River, removing many of the thornless shrubs. Two survivors available when the Scots Irish settlers arrived at the river were rusty black haw

(*Viburnum rufidulum*) and red haw (*Crataegus phaenopyrum*). The new settlers gathered the fruit of these shrubs in abundance. An observer noted that the Comanches advised howling like a wolf before eating red haws to avoid stomach cramps.[14]

The grasslands and herds of mustangs on the Southern Plains were two of the reasons that Comanches came to Texas. The most productive method for capturing these skittish animals was an adaptation of the antelope trap: a corral was built, perhaps in a creek bed, lined with heavy set posts and tall heaps of brush, and then brush wings were added to guide the driven horses into the trap. As the horses entered the trap, hidden riders dashed among them to rope one each, then throw the lassoed horse on its side. The Comanche then jumped from his horse to tie the captured mustang's hoofs together two by two. What followed was a gruesome choking treatment that some observers claimed rendered the horse tame and ready to ride in a single day.[15]

Personal honor was very important to a Comanche, and more respect was given when the capture was more difficult. If a firearm was available, the creasing technique might be used: a careful shot was placed through the muscle just below the mane. If the vertebrae were untouched, the horse would fall to the ground, briefly paralyzed, just long enough to rope and tie the animal. Even more prestigious was the rare and dangerous technique of lassoing a horse on the open range: the Comanche would determine which horse would be vulnerable to capture in a high-speed chase because of a hard winter, pregnancy, overweight from summer grass, or just after a long drink of water. This well-studied disadvantaged mustang could be ridden down, lassoed, choked to exhaustion, and thrown on the ground for the taming ordeal. The horse would be taken to a place with deep sand or several feet of water for the first mounting.[16]

The most prestigious way to acquire wealth through horses was to steal them, the more danger, the more valor. A skillful and dangerous theft was like a coup, most honorably done if no one was killed. The theft of a guarded horse under impossible circumstances was an enhanced mark of honor. The more domesticated, grain-dependent equines taken from the Spanish were trophies often bred with mustangs that were good foragers that remained in prime condition on summer grass and winter browse. The Comanches were known as selective horse breeders, and they produced the valuable Appaloosa and other buffalo runners.[17] A man with plenty of horses was a wealthy, powerful man.

Comancheria: A Quarter Million Miles of Grassland

As mounted fighters, the Comanches built an empire powered by violence to establish their new territory known as Comancheria, which covered most of the Southern Plains. A powerful political influence and major economic player, they increased their population by switching from Shoshone matriarchy to male-centered polygamy, increased by kidnapped wives, their offspring, and slaves. Equestrianism changed the Shoshone-to-Comanche culture to favor intense individualism, acquisition of private property, and primitive capitalism.[18] The Comanche florescence was made possible by reinvention into a vibrant society that was also culturally regressive. In it, agriculture became a thing of the past, progress toward a more complex social organization was reversed, and there was increased hierarchy and diversification, which gave way to specialization. Soon after, the Comanches became crystallized into a predatory culture that would not be able to transition into coexistence with the next dominant migration of Euro-Americans.[19]

During the Comanches' 150 years of empire, they were creatively changing to find a balance between horse herding, bison hunting, and trade. The Southern Plains, especially as far south as Texas, provided an ecological advantage of rich pasture and mild winters for horse production. Horse herds in the Northern Plains were reduced during each winter, leaving the Plains tribes horse-poor and ready to trade. Comancheria contained around two million feral horses along with a steady supply from thievery and raids, which made the Comanches wealthy as they supplied the Great Plains, Mississippi River valley, and beyond.[20] To meet the growing demand, the iconic bison-hunting Comanches became pastoralists, arranging their frequent movements, sometimes more than once per week, to find good horse pasture and water. By the early nineteenth century, the quality of Comanche horses and mules, many of which they bred themselves, were considered superior to Spanish or Mexican stock. The Comanche population along with the voracious horse herds grew so fast that bands split into one hundred family-based rancherias.[21]

Bison Hunting—With a Little Help from Animal Allies

After market production there were plenty of horses for war and the running bison hunt. In prehorse days the bison hunt consumed much of their time in planning a community-orchestrated cliff-fall, whereas with horses the hunt had become an afternoon's excitement. Even with each lodge consuming about fifty bison per year, the herd numbers were not depleted, and the Comanches would soon be back to the horse herds.[22] A hunt leader was chosen to decide how to approach the herd, after that, individuals chose their tactics. At breakneck speed the rider caught up with the bison on the right side and released arrows into the animal's flank, aiming behind the last rib. Sometimes a lance was used for a more flamboyant kill.[23]

The Comanches used a short, recurved bow (easily managed while on horseback) made from the prized bois d'arc tree (*Maclura pomifera*). Some of these bows were able to draw seventy pounds and were sinew-backed or even reinforced with horn capable of shooting an arrow completely through a buffalo or a man, as a Paluxy Creek settler discovered in 1871.[24] The bowstrings were bison or deer sinew that had been separated into dental-floss-like strands that were twisted and glued. The toughest arrows were made from shock-resistant dogwood (*Cornus drummondii*) and tipped with metal trade points. Arrows were coded with the hunter's identity so he could allocate the meat as he chose, which always included families without hunters.[25] Firearms could not be reloaded on a running horse and were rarely used.[26]

The bison herds were usually reported by scouts, but the Comanches lived in an enchanted world where assistance from animals was never a surprise. Horned toads (*Phrynosoma cornutum*), whose name translated as "asking about the buffalo," were said to have run in the direction of a herd.[27] But it was the raven (*Corvus corax*) in the north and the crow (*Corvus brachyrhynchos*) in Texas that offered more immediate information. These "wolf-birds" have had a symbiotic relationship with wolves and coyotes since their earliest evolution because the birds were unable to kill or open the carcass of larger animals. Crows have been observed micromanaging the hunt as coyotes move through the brush in pursuit of small prey.[28] Transferring this relationship to humans to collect butchering scraps was a natural adaptation. The Comanches took notice when one of these birds "flew over the camp four times, dipping his head and cawing" as a sure sign that a bison herd was spotted and the raven/crow should be followed.[29]

Comanche Captives—It Was Just Business

Cruelty was not the point of a Comanche raid, with the peculiar exception of acculturated captives whose cruelty was exceptional.[30] The purpose was to collect the livestock, kill the men, and take the women and children to be enslaved, sold, or adopted.[31] The Comanche strategy between the 1760s and early 1830s was not to exterminate; rather, the idea was to kill as few people as possible and leave enough farmers and livestock for continued production. The Texas missions, settlements, and ranches were seen as an economic resource to be harvested from time to time.[32] From the earliest times the Comanches brought in outsiders and welcomed a continuing Shoshone migration into a growing society that increased through the 1700s while neighboring peoples declined.[33] Comanche population grew to forty thousand before a major smallpox epidemic in the 1780s, after which the numbers averaged between twenty and thirty thousand for the next sixty years.[34]

After the borders of Comancheria were secure, raids were mostly about livestock, with looting and revenge expeditions less frequent. Deliberate abduction of others for assimilation to recover losses was not usually the primary motivation for raids. Encounters that involved kidnapping were the most dangerous, costing more Comanche lives than could be replaced by assimilation.[35] Many captives were seen as commodities to be illegally sold in the Spanish and later Mexican markets, and after 1847 to the Utes who delivered them to the Mormon Indigenous slave market in Utah.[36] For those captives who became candidates for resocialization there was a general cultural guideline from which an individual Comanche could vary. Rape, as part of the transitional ritual, was condemned, but if the ordeal turned deadly and vengeful, that was an individual's choice. Children held in anticipatory wifehood or given directly to the protection of a family, as with Cynthia Ann Parker, could bypass the terrifying initiation.[37]

Much like the method of breaking horses in the transitional ritual, the captive was made to feel powerless and dependent through various degrees of torture. The ordeal was more intense for adults whose ability to resocialize and conform to Comanche expectations was already suspect. For these initiates the treatment could involve public beating, knifing, and firebrands as the captive was forced to scream and dance to the screams of his tormentors. This treatment was often fatal, or the captive might be released unharmed.[38] The successful candidate suffered a social death from which a Comanche would be born. Afterward, the initiate was treated kindly and given as a slave to a family, the entry level to Comanche society from which any other level became a possibility.[39]

Comanche Women: Passive Aggressive

Women might become warriors as feared as any man,[40] but they typically became wives expected to maintain the camp, tan hides, and have babies. Comanche men were generally not sympathetic to the suffering endured by the women. Some of the wives were from the matriarchal Shoshone motherland who had very different ideas about a woman's place in society, and resocialized captive wives were often equally unhappy. If they survived childbirth, Comanche mothers nursed the infants into the fourth year, creating a very close bond that was abruptly broken when the sons turned to the company of men. It must have been a painful experience to be subject to abuse from all men, including one's male children.[41]

Frostweed, squaw weed (*Verbesina virginica*) is a course, upright perennial growing to five feet. A Spanish name, *lengua de vaca* (cow's tongue) is descriptive of the shape and texture of the leaves. Blooming in October, the flowers are white, occurring in a cluster at the top of the stem. It is found in shaded areas along the banks of the Bosque River. The name *frostweed* refers to the freezing ooze that can occur during cold weather. Some call it "tickweed" because ticks are said to cluster near the plant; it has been called "richweed" because it only grows in deep rich soil. Indigenous peoples used the plant as an abortifacient and gynecological aid for postpartum issues.

Geyata Ajilvsgi, *Wildflowers of Texas* (Bryan, TX: Shearer Publishing, 1984); Matt Warnock Turner, *Remarkable Plants of Texas: Uncommon Accounts of Common Natives* (Austin: University of Texas Press, 2009). Illustrated by Jill Alford.

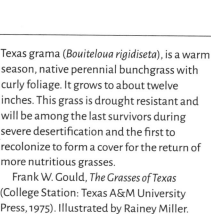

Texas grama (*Bouiteloua rigidiseta*), is a warm season, native perennial bunchgrass with curly foliage. It grows to about twelve inches. This grass is drought resistant and will be among the last survivors during severe desertification and the first to recolonize to form a cover for the return of more nutritious grasses.
Frank W. Gould, *The Grasses of Texas* (College Station: Texas A&M University Press, 1975). Illustrated by Rainey Miller.

Comanche women were not the servile procreators that their husbands assumed; the average birthrate of two children in a woman's lifetime was by design. Many women became sterile by choice.[42] An herb that Comanche women were known to have used is plentiful along the Bosque River, puccoon (*Lithospermum incisum*). The root of the puccoon was made into a tea that if used long enough would render permanent sterility.[43] If an unwanted conception had already occurred, Comanche women were known to beat their stomachs with stones to induce abortion, whereas women in the know might have used another Bosque River herb, frostweed (*Verbesinia virginica*), to terminate pregnancy in a less violent manner.[44]

Noting the absence of any disfigurement among the Comanches, observers wondered if selective infanticide had occurred. The chances of surviving to maturity were further tested by a life on horseback. Beginning in early youth, boys and girls were assigned up to 150 horses to herd and protect, changing pasture as needed, and stripping cottonwood bark as supplemental forage in winter.[45] After generations of forced marriages with Spanish, Mexican, and Indigenous captives, adult Comanches had become taller and heavier than the ancestral Shoshones or Hispanics.[46] While growing up managing the pastoral side of Comanche life, the youth were expected to be warriors in training.

Unsurpassed Mounted Fighting

Boys looked forward to owning their own war horse. Constant training placed Comanches among the finest horsemen in the history of the world. Central to Comanche training was their innovation of the loop, braided from the horse's mane at the neck, from which a leaning rider could pick up larger and larger objects. Soon these objects were picked up at a full gallop so that later a fallen warrior could be rescued in this manner. The neck loop and a second one were used for an ingenious technique of swinging to the side of the horse and releasing arrows from under the neck of the running animal at a frustrated enemy. A playful yet deadly maneuver was to vault around the saddle pad to face a pursuing enemy, then shoot arrows directly at him.[47]

Scots Irish migrants along the Bosque River described an early 1860s acrobatic display of horsemanship while chasing Comanches who had stolen horses. The Comanches were pressed hard by a large group of settlers, and the Indians had to abandon their herd and rush for the timberline. One Comanche fell behind the other riders, allowing their escape by distracting the pursuers. Riding bareback, he would flip around and sit facing the settlers while firing arrows to slow them down. The rider would then vault forward to gain some distance. The last time he swung around, he was near the timberline. Then he was struck from his horse by a limb, and, unable to find his bow, the Comanche pulled two arrows to make his stand until the pursuers surrounded him and shot him in the back.[48]

When on the offense the Comanches preferred a surprise attack. They would form a wedge that would soon split to enclose the enemy. Then the attack would be accompanied by disorienting yells.[49] The warriors would circle until the victims fired their muskets and then charge before the men could reload, discharging arrows at close range.[50] If the encounter involved captives, horses, and plunder, the Comanches could travel up to seventy miles a day, constantly changing horses. If the fleeing Comanches were overtaken, they were known to turn and encircle the pursuers and "charge them to the teeth."[51] Comanches were rarely known to surrender, and if one were overwhelmed he might dismount and take off his moccasins to signify his last stand.[52]

Near Dublin, Texas, in 1871 there was an exception to the Comanches' refusal to surrender. As three children were washing dishes, a hunger-crazed Comanche rushed in, scattered the children, and began scooping beans from a dish and "devouring them like a wild beast." A humane Baptist preacher, Ruben Ross, thereafter known as "Comanche Rube," arrived to see the man cross his arms over his chest, a universal declaration of friendship. He had lost his bow and horse and had been wandering along Armstrong Creek for days. The Comanche was cared for in Dublin until he was able to travel. After conversing with someone, probably in Spanish, he promised that there would be no more raids in or around Dublin. The promise was kept.[53]

Volcanic Eruptions Brought Americans Westward

Following a series of unfortunate though distant volcanic eruptions, the Comanches were about to be crowded away from their winter camps along the Bosque River. It was the last decades of the Little Ice Age, and the years between 1810 and 1819 were probably the coldest in five hundred years.[54] Travelers in 1872 recorded the eerie sight of dead mesquite trees in North Texas that may have been killed during the most intense cold after the cataclysmic 1815 Tambora eruption in Indonesia.[55] Tambora was the worst of a series of great eruptions that occurred between 1809 and 1815 and that loaded tons of volcanic aerosols into the stratosphere.[56] Confusingly, the 1815–16 "year without a summer" was not recorded in any of the Plains people's winter counts. The absence of mnemonic glyphs for these years has been interpreted as either a winter too disruptive for the hide documents to survive or the Great Plains experienced more moderate weather.[57]

Plenty of farmers were ready to give up in New England, an area that had suffered epidemics in 1813 and 1814, leading up to the hard 1816 volcanic winter. The way west was cleared by the victorious conclusion of the War of 1812 in 1815, which brought ethnic cleansing of Indians followed by a land rush into Ohio, Indiana, and Illinois that pushed Indigenous Americans and others toward Texas.[58] Anglo-Americans entered Texas from Louisiana when Napoleon Bonaparte sold the territory in 1803, where Philip Nolan had already opened the door in 1800 by successfully capturing feral horses and beginning the trade with the Wichitas and Comanches for much-needed firearms.[59] At first the Comanches were pleased with access to manufactured goods unavailable from a weakening Spanish presence in Texas.[60]

The same string of volcanic eruptions that began the Anglo-American migration westward neutralized Spain's control in Texas. In southern Mexico, some of the local elites, the American-born Spaniards, and the Creoles were beginning to press for more access to political power. Before the Creoles could act, another demographic rose in revolt. Volcanic-induced chill caused crop failures and created near-starvation conditions among the Indians and mestizos. This catastrophe created a desperation that Father Miguel Hidalgo tapped into in 1810.[61] Hidalgo's revolt horrified class-oriented Creoles, who refused to join the revolt, and Hidalgo was defeated and killed four months later. Hidalgo partisans and several semi-outlaw bands kept Mexico in chaos until less radical Creoles won independence from Spain in 1821.[62] The Mexican revolution left Texas decentralized, defunded, and unable to maintain status quo relations with the Comanches.[63]

The newly independent government of Mexico realized that Texas was nearing collapse under Comanche pressure, and the growing stream of Anglo-Americans flowing in from the border areas was recognized as an even greater threat. The response was Comanche appeasement to redirect their raiding activity toward the Anglo immigrants. The Comanches stalled for a year and then signed a treaty in 1822, with an understanding that they would report illegal Anglo intruders.[64] Comanche and Mexican fears came true in the 1820s when the nearly 17,000 Indigenous Texans and the Tejano population of two to six thousand were losing their dominance. The five thousand Penateka Comanches that frequented the Bosque River continued to outnumber intruders for a time, because the Anglo-Americans were concentrated along the coast.[65]

It was another result of the Tambora eruption that opened the doors wide for Anglo immigration. The first sustained United States economic depression was initiated by the market reaction to "famine-struck" western agricultural investment by East Coast grain speculators after the severe winter of 1816. When grain production came back strong in Western Europe and America, the huge investments collapsed and took the country and Moses Austin down with the Panic of 1819.[66] Nearly broke, bilingual

Austin traveled in late 1820 to San Antonio and began negotiations that ended in an endorsement on January 17, 1821, to bring three hundred Catholic-American Anglo families to Texas. Soon after, Moses Austin died and Stephen F. Austin, equally in financial distress, took up his father's project. Austin's grant was finally approved in January 1823, and his finances blossomed with a $60 fee from each family and twelve and a half cents per acre on the thousands of acres granted.[67] Added to this influx of planter-oriented Anglo-Americans were the more adventurous Scots Irish illegal immigrants from the upper South.[68]

Manifest Destiny: License to Kill

The Scots Irish frontiersmen had developed their ideal settlement pattern of scattered, isolated farmsteads, loosely grouped in neighborhoods three to six miles apart. A widely shared slogan was "no man should live so near another as to hear his dog bark." The Scots Irish brought their brand of militant Christianity that had taught them to believe in "vengeance and destruction of enemies.[69] The one thing the Anglo and Scots Irish Americans agreed on was that the Indigenous people of Texas would have to go, and braced, with cultural bigotry, God's will and genocidal ethnic cleansing became the guiding principles.[70] Playing right into this frontier attitude, the Comanches provided proof of concept when they began to raid Anglo livestock. Mexican officials had become too impoverished to continue with either diplomatic or military resistance to the Comanches.[71]

The Euro-American traders set up the Red River market depot for livestock captured from Tejanos and Anglos;[72] it delivered the only dependable manufactured goods, with the occasional firearm.[73] Comancheria contained an endless supply of feral horses, but breaking them took time. Stolen animals accrued prestige in their theft and raised the price for domesticated horses. Many of the captured feral or bred horses were far too valuable to sell or trade because the Comanches carefully trained them for bison hunting, war, and pulling a travois. They also traded mules valued in the Deep South market because of their heat tolerance and stamina. As the hide market dried up, mules became the specialty item that the Comanches bred from female donkeys, called jennets, brought back from their raids into Chihuahua and New Mexico to be mated with male horses.[74] It was simply more convenient to steal the livestock from the Anglos.

In 1825 Comanche raids intensified in northern Mexico on Tejanos and Anglos after Mexico City was unable to fund the annual peace maintenance through gift-giving. To the recent Anglo immigrants, the custom looked more like tribute or bribery, and they often refused, giving great insult. To the Comanches gift-giving turned strangers into friends, and these transactions came with obligations of honest and fair treatment to the parties involved. The gifts were allocated by the headmen to young Comanche men who had behaved responsibly over the last year, rewarding and honoring socialization. When no gifts were given the result was usually violence. In 1824 the governor of the recently unified states of Texas and Coahuila warned that the lack of gifts was about to bring a "total collapse of the peace."[75] The Comanche raids of 1825 were carried out by a mix of ethnicities: disobedient Comanche and Wichita youths, Tejanos, *ladrones* (thugs), and Anglo traders. Austin's partner, Baron de Bastrop, described the raiders as Mexicans and renegades rather than Comanches. The raiders behaved with uncharacteristic cruelty, not constrained by honor or custom; they didn't stop with taking captives and plunder, they went on to rape, and least honorable—to murder by ambush as well.[76]

These raids prompted Mexican officials to abandon Spanish policy in favor of Stephen Austin's notions of ethnic cleansing. In Mexico City, the response was to make a desperate gamble in 1825 to open the doors to immigrants in the forlorn hope that the Anglos would provide a barrier against the Comanches and become assimilated into Mexican culture.[77] The next year a disgruntled empresario, Haden Edwards, and some of the Cherokees in East Texas declared a new state called Freedonia. The revolution lasted only about a month and served as a loyalty test for everybody in Texas, causing the Anglo empresarios, eastern Native immigrants, and smaller Indigenous groups to declare support for Mexico. Mexican officials took advantage of these loyalty professions to sign up the various tribes to go against the angry norteños exiled from their homes in Waco and South Texas.[78]

The more responsible elements among the norteños that lived in exile on the upper Brazos and Trinity Rivers asked for peace. The resulting momentary treaty signed at Comanche Peak by Penateka Comanches, Caddos, and Wichitas was known as "the Peace of 1827." It was heralded in North American newspapers and accelerated the land rush to Texas.[79] In the fall of 1830 General Mier y Terán arrived with troops from Central Mexico and attacked the villages and camps of the Indigenous peoples who had signed the Comanche Peak treaty. First, he destroyed the rebuilt Waco/Tawakoni village on the San Gabriel River and the villagers' crops, then he destroyed the rebuilt Waco camp, and in November oversaw a much bloodier assault on a Wichita village in which the leading Penetaka Comanche voice for peace, Barbaquista, was killed after a hard fight. The Mexican, Anglo, and Cherokee campaigns of 1828–31 moved the Wichitas into permanent refugee status in North Texas.[80] After this time the Penetaka Comanches were without strong chiefs to direct the young men on the common good for the band, so unregulated raiding was then focused on status and individual wealth.[81]

Comanche political structure further weakened when a new round of smallpox was introduced by Anglo traders into Indigenous communities in January 1831.[82] This epidemic first appeared in Matamoros and Goliad before spreading to the Comanche, where the mortality rate was unknown. Early experiments in smallpox vaccination that year led San Antonio to try compulsory immunization of all children,[83] and the US government did successfully vaccinate some Eastern Plains peoples.[84] Perhaps the most destructive round of smallpox was the one mentioned in the last chapter when in September 1835, Colonel John H. Moore's adjunct, James Neil, infected a captive with an especially lethal type of smallpox.[85] The Comanches gave shelter to the fleeing Indians and then carried the virus northward, killing countless Natives in the Great Plains smallpox epidemic of 1836–38.[86]

Parker's Fort—Someone Left the Gate Open

There were other Comanche bands not yet infected with the virus who were outraged that the 1835 militia units burned, pillaged, and raped their way through agricultural Native villages along the central Trinity River. The Kichais, Caddos, and Wichitas were their trade partners and had to be avenged. The genocidal Ranger companies under John Moore, Robert Coleman, and Edward Burleson gathered at Silas M. Parker's stockade, which had just been built in 1834. Further raids, mostly for plunder, especially livestock, were conducted from Parker's Fort.[87] The next spring, Parker's Fort would become a target for revenge.

John Parker of Parker's Fort was a fiery fundamentalist Baptist preacher who cooperated with Comanches in the stolen horse trade. When he cheated them out of their share on a deal it was a more serious matter than he could have known.[88] Young men from the Wichita, Caddo, and other tribes, angry over the brutal attacks and loss of their homeland, were more than ready to join elements of the Noconi Comanche band, and even West Texas Kiowas were gathering for a revenge raid. The war band probably came together on the heavily wooded headwaters of the Bosque River and then followed the often-used river route toward Waco, cutting eastward to cross Tehuacana Creek and on to Fort Parker near the Navasota River.[89] The thirty-eight settlers at the fort were just beginning to relax after hearing about the Runaway Scrape and the Battle of San Jacinto that had removed the threat of the Mexican army.[90] When the war band of around one hundred arrived on the morning of May 19, 1836, most of the men were up to a mile away in their cornfields, and the handful inside the fort had left the gate open. A white flag deception was used to get close enough to ride inside the stockade, and during the following chaos five settlers were killed and another five were carried back up the Bosque River in captivity.[91]

The women were soon ransomed while Lucy Parker's children, John and Cynthia Ann, were resocialized as Comanches among whom they lived for decades. It was Cynthia Ann's story that became the most remembered captivity narrative among many since the 1600s, and it was appropriated to reinforce the savage versus civilized narrative.[92] The immediate purpose of these traumatic tales, reinforced with fictive cruelty, was to generate sensation and book sales, even though the captives often denied that torture and rape were part of their experience. The result was to draw attention away from Anglo excesses like the tens of thousands of deaths caused by biological warfare carried out by Texas Rangers and the Indigenous people killed and their villages burned a few months earlier.[93] The Euro-American hostilities of 1836 were called battles, while the five people killed at Fort Parker were lamented as a massacre.

Lamar—An Avatar of Manifest Destiny

In 1838 President Sam Houston, of the Republic of Texas, lost the election to Mirabeau B. Lamar, who announced an end to Houston's ignored peaceful policies. In his inaugural speech, Lamar declared war on all eastern immigrant Native and Indigenous peoples and referred to them as "red niggers" and "wild cannibals of the woods," probably quoting from the language of the captive narratives. Lamar believed, like most southern men, that the Anglos were destined to inherit the earth. When the Texas Congress voted for a million-dollar war appropriation in January 1839 and raised an army of one thousand, two thousand showed up.[94] Lamar's men hunted down Comanche camps in the west but were never able to strike a decisive blow. The army had better luck in East Texas in July when they burned all of the villages of the Cherokees, Delawares, Shawnees, Caddoans, Kickapoos, Creeks (Muscogees), Seminoles, and other refugee peoples.[95] The Cherokees who had helped carry out Austin's ethnic cleansing a decade before were now the victims of Lamar's genocidal purge. The Cherokees fought a disastrous rear-guard action to cover the people's escape to Oklahoma in July of 1839. Some of these refugee Indians tried to seek sanctuary in Mexico, but Lamar's army caught up with them, killed the men, and brought back the women and children to an unknown fate.[96]

In 1840 after the 1836–38 smallpox epidemic had abated and there was a break in hostilities, Penateka Comanches sent representatives to San Antonio to ask for peace. Anglo-Texas officials were instructed by the Lamar administration to impress on the Comanches that they would return all captives, avoid the settlements, and submit to Texas officials to "dictate the conditions of their residence."[97] Three companies of Lamar's soldiers arrived at San Antonio with plans to capture and hold the chiefs in exchange for captives. Sixty-five Comanches rode into San Antonio unaware of these plans or the unacceptable declarations with their families and settled in the plaza. Twelve Penateka leaders met with Colonel William Fisher and officials in a building known later as the Council House, where the rules of negotiation were suspended.[98]

The Penateka spokesman was Chief Muguara, who with the other leaders sat on the floor facing the Anglo officials; neither side knew enough about the other to avoid a violent outcome. The Comanches had no idea what a blunder they had made when they brought in Matilda Lockhart as their only captive. She had been horribly tortured by having her nose burned away. The Comanches may have considered this pitiful captive as a stratagem to increase urgency in future trade for captives. If they had learned anything about Anglo values and customs they would have realized that bringing in this particular captive was a fatal provocation.[99] The first question the commissioners asked was about the other captives, when Muguara responded that they were held by different bands but could be delivered for a price. If the Anglos had any knowledge of Comanches, they would have known that taking the roomful of chiefs captive would be impossible without violence. In response to Muguara's words, Colonel Fisher ordered the waiting soldiers into the room as the very nervous translator explained to the Comanches that they were now prisoners until all of the captives were brought in. There was only a moment of shock before the Penatekas leaped to their feet and tried to fight their way out the door.[100]

As the surviving Comanche headmen rushed out the Council House door they were met by a volley fired by the soldiers positioned outside, which killed two Anglos as well as more Indians. As the remaining Comanches ran for the river, the residents of San Antonio accompanied the soldiers in running down and killing or capturing all sixty-five of the negotiating party and their families.[101] A visiting Russian naturalist selected several Comanches' heads and the bodies of a man and a woman to boil down to extract their skeletons. The scientist was arrested and fined not for his gruesome bone collecting but because he had thrown the human remains in the city's water supply, the San Antonio River.[102] One of the captured Comanche women was given a horse and the message that unless all of the Anglo captives were brought in, the jailed Penateka women would be killed. She never returned and the remaining Comanche women either escaped or were later exchanged.[103]

After several exchanges were made, another dozen Anglo captives were tortured to death and a major revenge raid was planned by the rising Penateka Chief Buffalo Hump in response to the Council House tragedy.[104] In July 1840, a force of five hundred Comanches and their families raided to the Gulf Coast, killing nearly thirty residents and burning the port city of Linnville. On the return trip, the Penatekas were intercepted on August 12 at Plum Creek, a branch of the San Marcos River, by the Anglo-Texan army and local militia with Tonkawa auxiliaries. The Penatekas were able to protect the women and children but lost one hundred of their people before escaping northward into Comancheria.[105]

In October Colonel John Moore was sent up the Colorado River and ordered to attack the first Comanches he found. This time the Indigenous allies were the Lipans, who located a Comanche camp, and Colonel Moore spent the night preparing for the morning assault. Moore placed his riflemen across the Colorado River to shoot survivors as they swam to safety. The Rangers and Lipans "slaughtered people in their beds," killing 130 men, women, and children, after which the Penatekas moved north of the Red River.[106] Then Lamar's army made a sweep through the Cross Timbers to Village Creek (Arlington) where they were repulsed by the agricultural

Natives. General Edward Tarrant renewed the depredations the following May of 1841 in the so-called Battle of Village Creek when his men burned several villages as people scattered for the brush.[107] Lamar's harsh policy opened the Dallas–Fort Worth area to Anglo settlement, while the stunned Indians avoided the settlements.

The war had exhausted Anglo-Texan patience with Lamar. He had sent a disastrous expedition to Santa Fe in which all of the Texans were captured; he had also antagonized the pro-annexation voters and alienated the planter-merchant elite who were outraged at the massive debt since they were the ones who paid most of the taxes.[108] By 1842 there were so few Natives left in Texas that Ranger units could no longer collect their pay from the thousands of dollars in stolen livestock and Indian goods, which ended one of the "most lucrative examples of looting in the history of the American frontier."[109] Sam Houston's reelection campaign was based on a return to peaceful, inexpensive coexistence. His Indian policies cost $200,000 while Lamar's extravagant expansionist policies had run up a two-and-one-half-million-dollar tab. Houston was inaugurated in December of 1844 and reinstated his peaceful policies. Caddos, Wichitas, and Penatekas Comanches returned south of the Red River and remained peaceful until the end of the Republic of Texas.[110]

The Plum Creek engagement was the last time that the Comanches would meet large numbers of heavily armed Anglo-Texans in an open fight; stealthy raids were much safer thereafter. Samuel Colt's 1836 revolving firearm was beginning to enter Texas by 1840, where it was known as the Texas Patterson. These five-chambered pistols, perhaps first used at Plum Creek, became famous when on June 8, 1844, Major John "Jack" Hays, Samuel H. Walker, and fifteen Rangers attacked a hunting party of sixty Yamparika Comanches, Kiowas, and a few Shoshones, most of which had never fought Anglos. The Comanches rallied and faced the heavily armed Rangers and lost twenty-three dead and as many wounded.[111] The "Walker's Creek Battle" was depicted on the cylinder of one hundred of the new, deadlier Colt pistols developed in 1847.[112]

The Walker Colt and More Diseases

Samuel Walker traveled to the Colt factory in Connecticut to advise Samuel Colt on the design for the Walker Colt, which became the most powerful handgun ever used by the US military. It had six chambers that loaded twice the powder as the earlier models and fired a .44 caliber ball with the muzzle velocity of a modern .357, with a cylinder that could be exchanged for another fully loaded one while on horseback. The mounted Anglo was now more than the equal of the Comanche archer, because this pistol shattered shields and could kill at one hundred yards.[113] This weapon ended Native resistance, except for young men still trying to achieve honor, and called down the wrath of settlers.

In 1844 the bands still totaled around fifteen thousand Comanches, eighty percent fewer than the peak years in the 1780s.[114] Becoming demoralized, Texas Natives attended talks for the next couple of years leading up to the US war with Mexico. However, the Comanches lost people with every encounter with Anglos, which was not the main reason for their rapid population collapse. After 1840 increasing numbers of Euro-Americans brought more diseases, accelerating the decline. For example, from 1846 through 1847, measles and whooping cough entered the Plains from the east. Mosquito-borne yellow fever was a danger anytime people traveled too far south.[115] The most dangerous disease during this time was one last ill-effect from the Tambora eruption of 1815, which traveled a long way to reach the Bosque River.

A cholera strain was brewed in the Bay of Bengal as a weather anomaly destroyed crops and sent refugees into a densely crowded chaos. The result was a new strain of cholera that killed ten thousand people in the first two weeks. British troops helped the new strain become the world's first pandemic as it spread around the world,[116] landing on the Texas coast in 1846, where it killed so many people in Indianola in 1846 that the bodies in the streets couldn't be buried for days.[117] By 1849 cholera was infecting waterholes and streams in the Great Plains, killing thousands of forty-niners. A pan-ethnic Sun Dance sponsored by the Kiowas attracted Comanches and other Plains peoples from miles around, but cholera broke out and killed the Natives before any symptoms were presented. Those who attended suffered 50 percent fatalities, and only scattering to freshwater camps saved thousands more lives.[118] In 1850–51 there were influenza and smallpox epidemics that killed mostly children, followed by another round of smallpox in 1856–57;[119] and venereal diseases were a constant horror.[120]

10

Conquest and Commodity

Many young Penateka Comanche men turned to whiskey, contributing to the collapse of the Bosque River's most frequent campers.¹ The source of the alcohol was most likely the new Bosque River residents. A community began after the Texas revolutionary war as a haven for Delaware and Shawnee leaders as a reward for their service as interpreters and diplomats with the Comanches and as auxiliaries in Texas's war for independence. Sam Houston gave land on the Bosque River to twenty families, relatives, and friends of the tireless advocates of peace including John Conner, Bill Shaw, and Jim Shaw.² The location of this land grant is still unknown. It was described as "a heavily wooded spot untouched by Anglo settlement on the Bosque River."³ The Delawares and Shawnees had endured two hundred years of useless treaties and forced removals and had acquired much of the Euro-American lifestyle, including clothing, log cabins, and a realistic mindset.

Perhaps the most outstanding personality in this short-lived community was John Conner, whose father was a successful trader in Ohio. His mother was an Anglo woman captured by the Shawnees at an early age, and after recapture he was raised as John Conner. Conner lived among the Delaware people and married the daughter of a prominent chief. Around 1820 he headed west, driven by what he later called an intense "wanderlust." Conner traveled alone and on foot to the Pacific Ocean and then down to Mexico in his early twenties. He traveled to Mexico City and wandered around northern Mexico for three years until moving to Texas and joining the Delawares who had migrated in from the crowded East.⁴

There were only about five hundred Delawares in Texas when Sam Houston enlisted some of them in a brigade to help in the Texan war for independence. Houston later referred to them as "the most efficient auxiliary corps to the main army."[5] Delaware men were well-off financially, serving as guides, translators, skilled hunters and trappers, and traders, often traveling long distances to hunt and trap. Ranger captain John C. Hays was invited to travel on foot on an expedition from the Western Cross Timbers to the Pecos River with seventeen Delawares in the fall of 1838. One of their party was killed by Comanches, and the young Ranger was amazed to see the Delawares run for hours until they caught up with those Indians on the Rio Grande after a few days; they killed almost all of them in a morning attack.[6]

John Conner, as Houston's aide, was present at nearly all negotiations and treaties during the 1830s and 1840s, not just as an interpreter but also as a persuasive diplomat to the Comanches, whom he successfully convinced to attend treaty talks. Negotiating with the Comanches had become more difficult after the 1840 Council House treachery. Conner was described as "justly renowned as having a more minute and extensive knowledge of the continent than any other man," making him the ideal choice for frequent missions into Mexico and as a guide for the Chihuahua–El Paso expedition in 1848. He was compensated with a league of land (4,428 acres) granted by the Texas legislature in 1853, the only native American to receive such a grant.[7] By 1857 Conner was dividing his time between Texas and Kansas when he was authorized by the US Department of Indian Affairs to become the principal chief of the Delaware Nation.[8]

Two other Delawares present in the Bosque River village were the brothers Jim (Bear Head) and Bill Shaw (Tall Man). Jim Shaw was described by William B. Parker as "the finest specimen of the Indian I saw during the trip, about fifty years old, full six feet six in height, as straight as an arrow, with a sinewy, muscular frame, large head, . . . his countenance indicative of the true friend and dangerous enemy."[9] Jim Shaw was especially distinguished in Indian-Anglo relations, and before arriving in Texas in 1841 he spent time along the Red River learning Native languages. Shaw was well-known in Indian Territory by government officials as a useful diplomat, a successful trader, and a man of superior intelligence. The US Indian agent recommended him to Houston, who employed him to help carry out his peace policy.[10] Shaw was able to win the trust and friendship of several Penateka Comanche leaders including Buffalo Hump. In 1846–47 he was able to use his Comanche relations to guide German colonists under John O. Meusebach to a successful colony that developed a lasting friendship with the Comanches.[11] In 1854 the Shaw brothers and Colonel Robert E. Lee surveyed the site of the Upper Brazos Indian Reservation. Jim Shaw died in 1858 from a fall while repairing the roof of his log cabin.[12]

Many of the likely residents and frequent visitors in this Bosque community were larger-than-life frontier characters, like Scots Cherokee Jesse Chisholm. Chisholm was born in Tennessee around 1805, then ended up in Texas around 1836 as a successful trader who soon enlisted in the service of the Republic of Texas. Chisholm had learned more than a dozen languages during his years as a Plains trader and was an invaluable interpreter at treaty councils in Texas, Indian Territory, and Kansas.[13] Chisholm had the respect of many Comanche leaders and was sent by Houston in 1842 to persuade those headmen to come to the council. The Comanches refused until some of their captives were returned. So Houston wrote to Lamar's scourge, John H. Moore, demanding that the colonel send any Comanche prisoners taken during his raids. Tragically, the young women had already been sold by the Rangers into slavery among Anglo-Texans.[14] To continue diplomatic efforts among the Comanches, Chisholm was invited to add his cabin to the Bosque River grant.[15]

Jesse Chisholm and the Shaw brothers assisted in bringing Comanche and other leaders to the 1846 Treaty of Comanche Peak, and then Chisholm accompanied an Indian delegation to Washington, DC, where he translated for President James K. Polk. By the time of the Civil War, Chisholm had established a chain of trading posts from Texas to Kansas. During the Civil War, Chisholm remained neutral, providing livestock and interpretive services for both sides on the Southern Plains. And in 1864, as part of a government relocation project, Chisholm was made supervisor of fifteen hundred Wichitas whom he guided north of Oklahoma for their safety.[16] At his ranch in Wichita, Kansas, Chisholm had a herd of three thousand "Indian Cattle," a breed developed by the Cherokees from Spanish and English stock. It was during his care for the Wichitas, who were in his charge, that Chisholm cleared more trails into wagon roads to deliver Indian cattle to US forts. By the end of the war the wagon road, never used by Chisholm to herd cattle, became the Chisholm Trail.[17]

Not all of the likely Bosque River outstanding personalities were tireless workers for peace. Some, like the Afro-Delaware Jim Ned, and his brothers Jack and Joe Harry, had mixed reputations. Jim Ned's father had been a slave, and to conceal his ancestry he shaved his head and wore Delaware turbans or other headgear. He married a Delaware girl and was soon immersed in the culture. In his younger years, he was like the other Texas Delawares, a valuable interpreter, scout, and spy, but later devoted his efforts to stealing horses and causing trouble. Ned spent time among the Wichitas and Penateka Comanches and became familiar with their customs only to be exploited later. Jim Ned attracted renegades from the Texas frontier—sometimes numbering as many as 6,100—that traveled beyond agreed boundaries to hunt, raid, and trade.[18] For whatever reason, Jack Harry rushed into Waco to announce that two hundred Waco Indians were about to attack Torrey's Trading House and the neighboring settlements. Militias were called up, and there was much excitement for a time. Other rumors connected the brothers to Jim Ned's horse-stealing activities.[19] The renegade Delaware brothers alternated between selling whiskey, encouraging Texas Natives to violence, and guiding official government expeditions until the Natives were moved to Indian Territory.[20]

The Native immigrants from the east, Delawares, Shawnees, Seminoles, Cherokees, Caddos, and Kickapoos, were described as "decently clad" in imported, colorful textiles, turbans, and red sashes, and they kept themselves very clean. The women dressed simply in long dresses that were compared to ones worn by French country girls, adding native shawls and colored ribbons in their hair. The women wore silver rings and earrings of feathers and dried bird skulls.[21] By far the most colorful and flamboyant dresser who probably visited the Bosque River "town" was the already famous Coacoochee, Wild Cat, the Seminole war chief. Back in Florida, the Seminole leader had shown up for a series of surrender talks with none other than Lieutenant William Tecumseh Sherman in 1841, dressed as Hamlet after the chief murdered a traveling troupe of Shakespearean actors near St. Augustine.[21]

Wild Cat was born around 1810 to distinguished Seminole parents and was captured in 1837 along with Seminole leader Osceola, while under a flag of truce. He escaped prison and continued to resist the US campaign of conquest in Florida, frustrating the likes of Colonel Zachary Taylor by ambushing soldiers and then vanishing back into the swamp. The war cry of Wild Cat's Seminole warriors, *Hooah!*, was adopted by the US military[22] during the last campaigns of 1835–42.[23] Along with two hundred followers,

Wild Cat agreed to be taken to Indian Territory in 1841, from where he and his captains traveled to the Red River to meet with representatives of the US Army, Texans, Kickapoos, Caddos, and other Native peoples.[24] Wild Cat must have impressed the Indian agents because they escorted him into Texas to try his hand at peace talks with the Comanches.[25]

During this time Wild Cat and his associates most likely showed up at the Bosque River "town," as it was being called. He was described as "about forty years of age, slight, active, and well-proportioned, with an attractive and intelligent appearance," and was an alcoholic.[26] Gato del Monte, as he was sometimes called, had in mind something very different from what either the Comanches or the Indian agents were hoping for during the talks. In the Indian Territory, the Seminoles were forced to live amid their Muskogean forebears who regarded them as "so many dogs" whose African contingent were considered fair game to be kidnapped and sold into slavery in the Deep South. Wild Cat's dream was to gather all Texas Natives, free Blacks, and escaped slaves into a confederation to travel to and establish a colony in Mexico.[27] The Comanches were not interested, but several other refugee Native groups that Wild Cat met at places like the Bosque post were very interested.[28]

Among Wild Cat's captains was another larger-than-life character, John Horse, Caballo Juan, who was better known as Gopher John. Like many Seminoles, Gopher John was the son of an Afro-Native mother and an Indian father with a trace of Spanish ancestry. He was a "big, tall, fine-looking, ginger-colored man, with a proud carriage and renowned for his coolness and courage, and for his deadly accuracy with a rifle." The story of how Gopher John got his nickname touches on early nineteenth-century politics. In 1826, long-legged Afro-Seminole young John was told by an army officer that he would pay him twenty-five cents for each gopher tortoise (*Gopherus polyphemus*) that he could find because gopher tortoise was the officer's favorite dish. When the tortoises were turned over to the cook, she placed them in a box. For the rest of the day, John picked out the gopher tortoises and sold them over and over to the delighted officer. The incident made the newspaper and was chuckled about until David Crockett heard the story; Crockett was looking for a homespun story to add color to a speech he was preparing to win a seat in the Tennessee House of Representatives. Crockett altered the story he told as a time he was sitting in a bar paying for his drinks with the same coonskin over and over. The story went over so well that Crockett adapted the coonskin cap as part of his populist image.[29]

While Wild Cat was still fighting in the Seminole resistance, Gopher John was able to escape the Seminole Wars in 1838 because of his skills as an interpreter for the US Army. Until Wild Cat could join him, Gopher John traveled back and forth from Florida and Indian Territory getting hundreds of belligerents to surrender and helping many of them to relocate.[30] Gopher John shared Wild Cat's plans for a Mexican colony, and the ideas were shared during his promotional visits among Texas Indians. Officials in Texas were annoyed that Wild Cat used his eloquence to alarm the Natives about what their future was going to look like under the Anglo-Texans, and Wild Cat set an example to other Natives by refusing to smoke with Texans.[31]

The Bosque River sanctuary had become notorious as a source of liquor for the Comanches and other Native peoples, so when Texas was annexed by the United States in 1845, Houston's land grant was discarded and the "town" was closed.[32] The new Indian superintendent, Thomas G. Western, explained that because Anglo settlements were getting too close to the Bosque community, Natives were being attracted too far south. Jim Ned and the others left the Bosque River log cabin community under protest and moved their whiskey shops farther up the Brazos River to trade with Comanches and with other Natives in the Indian Territory.[33]

Genocide for Fun and Profit

Meanwhile, after the United States' war with Mexico, the dynamic leader Wild Cat had been recruiting dissatisfied Texas Indians and eastern refugee Natives along with free and escaped Afro-Americans for his Mexican colony scheme. In 1849 Wild Cat tried one more time to enlist Comanches, Caddos, and Wacos to join his exodus, explaining that in Mexico they could legally fight the Texans when they tried to raid the colony. The only people interested were a few hundred Kickapoos whose reputation as skillful fighters was on a par with the Seminoles.[34] Wild Cat and John Horse led separate groups of Seminoles and "Black Seminoles" on a yearlong trek to Mexico, camping for extended periods on the Llano River and Las Moras Springs near Brackettville.[35] They were shadowed by Creek and Anglo slave hunters, Texas Rangers, and Comanches, who had learned of the military nature of the migration and the plan to serve in the Mexican army, to block predatory raids into Mexico. Some Delawares and Shawnees, suspected of going to meet Wild Cat, were killed and members of a separate party of sixty Blacks were murdered and their families kept for ransom.[36]

In November of 1850, Wild Cat's group of three hundred arrived, "armed to the teeth," near Piedras Negras where Wild Cat (now Gato del Monte) was commissioned colonel, and John Horse (now Juan Caballo), was made captain in the Mexican army.[37] The Mexican government granted land, farm implements, and provisions to the Black Seminoles, Seminoles, and Kickapoos. Because of their reputations as fierce fighters and deadly marksmen, they were assigned to protect the northern Mexican frontier after planting their corn. Successful at repelling Comanche and Apache raiders, the colonists soon found that there were other hostile accomplices about to descend upon their villages.[38] Texas politicians took action, claiming that as many as four thousand slaves had fled to Mexico by 1855. The governor helped generate tales of Seminole invasions, which encouraged Texas Rangers, private Ranger companies, and other adventurers who hoped to profit from slave raids into Mexico to invade the area around Piedras Negras. Each time they were vigorously repulsed by veteran Seminole colonists.[39]

Texas Rangers had always regarded livestock, captives, and plunder from Indian villages as part of their payment for pursuing raiding Indians. The excitement of legal killing for profit rendered many of them unable to settle down to farm or ranch life. The result was "a large class of people ... who prefer the wild and indolent life of the volunteer ... [whose] one desire was to find and exterminate Indians."[40] This lasting frontier creed led to a culture of violence reinforced by captive narratives and newspapers. The sedentary tribes were desperate and unable to stay in one place long enough to bring their crops to maturity. They needed a safe place from roving Anglo predators and land-hungry soldier-settler encroachment.

The Texas Indian Reservations

A safe place, a reservation for Texas Natives, was the dream of Virginia-born Robert Simpson Neighbors, who would soon give his life in the service of these people. Arriving in Texas at nineteen in 1836, Neighbors served with distinction in the army of the Republic until his federal appointment as Indian agent in 1847.⁴¹ Neighbors worked for years to convince the Texas legislature to authorize the federal government to take control of twelve leagues of vacant land "for the use and benefit of the several tribes of Indians residing within the limits of Texas."⁴² In 1855, two reservations were opened, one at the confluence of the Brazos and Clear Fork Rivers for the near-one-thousand remnant agricultural peoples and another, forty miles away on the Clear Fork, for the few cooperating Comanches.⁴³ Before long the sedentary Natives had built more than a dozen log cabins and over a hundred traditional grass dome houses. Livestock was rail-fenced away from the corn and eight hundred peach trees. Neighbors saw himself as an agent of acculturation who would Americanize the Indians through the Christian religion, horse-drawn agriculture, and literacy. There were log buildings for agricultural instructors, teachers, schools, interpreters, and Indian agents. And the Natives could leave the reservation to hunt with permission.⁴⁴

Attracting Comanches to their reservation was more difficult. The Mexican-American War (1846–48) marked their apex of power and prosperity when they still believed themselves the most powerful people in the world. Even after constant epidemics, there were still around twenty thousand Comanches, giving them every reason to feel confident. But after the cholera epidemic of 1849, the Penatekas were "literally destroyed as a tribe" yet still unable to alter their raiding lifestyle.⁴⁵ Raiding into Mexico was part of the Comanche cycle for 150 years. Some who found the Texas frontier intolerable moved to Mexico, and under the leadership of a mysterious old woman, "the prophetess" Tave Pete,⁴⁶ signed a treaty with Chihuahua, promising to direct their raiding toward Nuevo Leon and Coahuila.⁴⁷ The northern bands and Buffalo Hump's Penatekas stayed out of Texas between raids. The remaining battered Penateka bands on the Texas frontier, led by Chiefs Sanaco and Ketumsee, during drought conditions cautiously showed up at the Comanche Reserve on the Clear Fork of the Brazos River since food was scarce during the drought.⁴⁸

As nomadic hunters, the southern Comanches struggled with the concept of becoming farmers living in permanent villages contained within boundaries. A core group of Penatekas were able to set aside traditional abhorrence to horticulture and adjusted to a life as farmers.⁴⁹ The Comanches on the Clear Fork Reservation amounted to around five hundred, mostly Penatekas. Thousands of others refused to enter the reservation; Buffalo Hump's Penatekas, most Kiowas, all of the Lipan

Apaches, and some bands of Wichitas and Tonkawas continued to resist and raid, confirming the Anglo narrative of the "depredating savage."⁵⁰

Stephenville: Anarchy and Hysteria

Meanwhile, at the same time that the reservations were being established, Anglo settlers were moving up the Bosque River, moving the frontier closer to the reservations. In 1853 Norwegian settlers started a community halfway up the Bosque River. The families of the men who died at the Alamo were awarded 4,409 acres; the John Blair family were recipients and decided to sell the survey located at the present site of Stephenville. In 1848 James Stephen bought the land and six years later traveled up the river from Waco to establish an unnamed slave family on the site to determine the disposition of the Natives. A year later the surveyor George B. Erath and Neil McLennan escorted the Stephen brothers, thirty families, and eighteen slaves to the Stephenville site in 1855 on the Bosque River. Almost all of the first log cabins were built around the town square amid heavy post oak timber.⁵¹

At the time Stephenville was the farthest west of Anglo settlements "on the waters of the Brazos," located on an old bison trail from Brownwood to Fort Worth, now Highway 377. The game was plentiful, with bison within three miles of town, and there were fish and large alligators in the Bosque River.⁵² After a couple of peaceful years trading with José María's friendly Anadarkos, who were permitted to hunt in the vicinity, there was an incident that Stephenville settlers thought was responsible for two years of violence along the frontier.

Red Jack was known to Bosque River residents as a decent man who never touched alcohol until the fall of 1857, when he left his rifle and bow at a pair of post oaks at the present Tarleton State University and rode into the Stephenville village. For some reason, the Anadarko bought a fifty-cent bottle of whiskey and became so drunk that he was asked to leave town. On his way back to his gear, he saw a log cabin and decided to try to ride his horse inside. No one was home except a terrified and sick mother and her sixteen-year-old son who snapped the capped but unloaded revolver chamber at Red Jack. Stupefied by drink, the Indian panicked and pulled his knife, causing the boy to fire a live round. Red Jack remained on his horse, falling off dead before reaching the tree and his gear, which wasn't found for two more years.

Because the Anadarkos had a fierce reputation as fighters, the terrified residents laid his body in state for as long as the smell allowed, then buried him.[53]

José María, known all over the frontier as a hard fighter but fair man, showed up in Stephenville with one hundred men to inquire about the death, and, while offered food, listened to the story and left.[54] Later that winter in 1857, Stephenville experienced its first theft of livestock, and, probably coincidentally, the whole line of the frontier from the Rio Grande to the Red River experienced a violent winter of "murders and robberies."[55] Violent attacks and responding hysteria by Anglos marked the years 1857–59 on the Texas frontier. Some of the raids were carried out by Northern Comanches from Indian Territory in Oklahoma; some few reservation Comanches probably joined them, yet in large part the raids were conducted by Anglos disguised as Indians.[56] As it turned out, the anarchy and hysteria on the frontier were engineered by a core group as a cover for genocide, larceny, and real estate.

The Texas legislature passed a bill on January 26, 1858, authorizing Governor Hardin Runnels to send Ranger Captain John "Rip" Ford one hundred more Rangers to block the Northern Comanches from raiding south of the Red River. Growing bored with watching the reservations, Ford decided to exceed his orders and search for Comanche camps above the Red River.[57] Ford was able to convince Neighbors to help José María (now spokesman for most reservation Natives) to recruit 109 of the sedentary Indians to serve as auxiliaries and prove their loyalty to the people of Texas. The combined Ranger-Native forces crossed the Red River and attacked a Comanche camp on Little Robe Creek, on May 12, 1858, killing seventy-six Comanches and capturing eighteen women and children. Among the Comanches killed was Chief Pohebits Quasho, Iron Jacket.[58]

The Ford–José María raid outraged the Comanches and led to more frontier violence, which was met by a second raid inside the Indian Territory. This time reservation Indians joined Major Earl Van Dorn, and elements of the Second US Cavalry, which crossed the Red River. Soon the reservation Indians reported a Comanche camp. It was Buffalo Hump's camp on Rush Creek, right where Lieutenant J. E. Powell instructed him to camp until the peace they had agreed on could be made official. Tragically, while Buffalo Hump was preparing to travel to Fort Arbuckle to make the treaty official, Van Dorn's cavalry charged through the village on October 1, 1858. Dozens more Comanches were killed, accelerating frontier violence further.[59]

The Choctaw Tom Massacre

Stephenville vigilante leader Peter Garland was one of many firebrands to accuse the reservation Indians of being the culprits to a frontier audience of recent, gullible immigrants to Texas. After the accusation Peter Garland gathered twenty Erath County residents on a mission to scout the Brazos Reserve, as many vigilante gangs were doing, hoping to find Indians off the reservation.[60] Garland's gang was turned away at the reservation, and on the return trip arrived at Golconda (later Palo Pinto) fifty miles north of Stephenville. There, Garland learned that a family of reservation Indians were camped nearby at the invitation of rancher C. C. Slaughter, with permission of Indian agent Shapley P. Ross, to hunt bear for the winter's cooking oil.[61] The people of Golconda explained that the Natives were friendly and should be left alone; perhaps they mentioned that the family patriarch, Choctaw Tom, had been a scout for Sam Houston and that two of the young men were with Van Dorn on his attack on Buffalo Hump's village in October.[62] Garland and his eager Indian hunters explained, deceptively, that they were starting back to the Bosque River.[63]

The vigilantes knew that no Anglo had been questioned about murders committed around the Brazos Reserve, so at dawn they crept up on Choctaw Tom's Caddo families and a few Anadarkos. Choctaw Tom had just purchased a wagon and had already started back to the reservation, leaving his family and friends asleep in camp. Choctaw Tom's wife was among seven people killed, mostly women and children in their beds; a nine-year-old girl survived by pushing a rifle away from her face as her thumb was shot off. Sixteen-year-old Samuel Stephen, whose father had not approved of the raid, was shot in the back of his head by his own men.[64] When Garland led his victorious men back to Stephenville, he announced, "We have opened the ball and others can dance to the music."[65]

That cryptic remark was taken then and now as an alarm of impending attack by José María's Caddos and Anadarkos. Many young men on the reservation wanted to do just that, but José María, as many times before, calmed the situation by promising the matter would be taken to the courts.[66] Garland and W. W. McNeill, another Stephenville provocateur, published their contrary version of events. They had been following the trail of horse thieves that led them to the Caddo camp where a "fierce and terrible battle" ensued and "we have no apology for what we have done."⁶⁷ In early January 1859, as many as three hundred hate- and fear-crazed settlers from nearby counties gathered in Stephenville in preparation for an attack on the Brazos Reserve. Not only did they endorse the Choctaw Tom massacre, but they also elected members of the Garland raid as their captains. Allison Nelson, a longtime provocateur, was elected commander of the vigilante army on the Bosque River.⁶⁸

The well-known Ranger and surveyor, George B. Erath, was among a committee chosen to travel to the reservation to determine whether or not an attack was imminent. The Indian agents had not yet returned from their Christmas leave in San Antonio, so Erath and the "commissioners" met with Captain T. N. Palmer of the Second Cavalry and others who tried to convince the vigilantes that the reservations were peaceful. Erath summed up the meeting: "more or less was said to little purpose." When the commission returned to Stephenville and made its report, most of the men went home, at least for a time.⁶⁹ In reference to Garland's band, celebrated as heroes along the Bosque River, western historian Hubert Howe Bancroft said the names of these men were doomed to "immortal infamy."⁷⁰

When word of the Choctaw Tom Massacre reached District Judge Nicholas Battle, he began preparations to prosecute Garland's vigilantes. Judge Battle mistakenly believed that the rage and hysteria in the frontier counties could be brought to reason if only the Garland gang could be apprehended and punished. At first, he tried to get officials in Palo Pinto County to indict the killers, but they would not, and instead they indicted José María for horse theft. Next, Judge Battle appointed E. J. Gurley of Waco as special prosecutor to gather evidence.⁷¹ Major Neighbors and one of the few Erath County men in possession of his reason, Joshua R. Carmack, made affidavits for the arrest of Garland's vigilantes. Then the judge issued writs to Ranger Captain Ford to arrest Garland and his men.⁷² Ford refused to arrest the vigilantes, even when ordered by Governor Runnels, and the frontier continued to slide into anarchy.⁷³

John R. Baylor: Frontier Demagogue

The best remembered and most gruesome instigator of the chaos that stained this period of history along the upper Bosque River was John R. Baylor, who directed characters like Peter Garland. Baylor had been appointed Indian agent to the Comanches on the Clear Fork reserve in 1855, and in 1857 was dismissed for feuding with his supervisor, the conscientious Robert Neighbors. Baylor claimed that Indian agents were operating a horse-thieving scheme in which they ordered the reservation Indians to steal horses from the frontier, which were delivered to Kansas markets.[74] Captain Ford investigated the charges and found that the hoofprints of the "relentless, merciless, and treacherous foe" were shod with iron shoes and the arrows were not even Comanche.[75] Edward Burleson Jr. reported that Baylor had asked him to help steal a neighbor's horses and then fabricate a trail to the Comanche reservation.[76] Others reported that Indian trails were usually hard to follow, but after some of these murderous raids the trail was so distinct that "an Easterner could follow . . . as there were blazed from mutilated bodies to the Clear Fork Reserve."[77]

In a supreme act of projection, Baylor was found to be guilty of the very scheme that he had worked so hard to spread during mass meetings and through his tasteless newspaper called *The Whiteman*.[78]

A letter was intercepted by one of Baylor's operatives that explained the stolen horse traffic to Kansas, implicating the for-profit demagoguery that had created the culture of violence on the frontier.[79] Many of the atrocities on the frontier were committed by Baylor's scoundrels disguised as Indians and the Patrick Murphy gang, who were implicated in the letter but considered themselves invulnerable.[80] Sam Houston looked into the matter and wrote to a friend that the frontier chaos originated through "the prejudices of evil and designing persons, who keep up the cry to enable them to better carry on their schemes of robbery and plunder."[81]

By May the charismatic Baylor and his core of desperadoes had convinced themselves and hundreds of others that the best policy for the Brazos Reserve Indians was not just removal from Texas but extermination.[82] Baylor and Garland gathered as many as five hundred vigilantes in Jacksboro under a banner presented by the local ladies: "Necessity knows no law"; they approached the Brazos Reserve on May 23, 1859.[83] Baylor divided his vigilante army into two halves, 250 men under Peter Garland tried unsuccessfully to capture the artillery at Camp Cooper,[84] while Baylor and the other 250 rode to the Brazos Reserve where José María and fifty fighters and the First Infantry under Captain J. B. Plummer were camped in waiting.[85] Baylor formed a line of battle and began a round of threatening discussions in which Baylor said he would regret killing US soldiers but he would if they resisted. With the reservation officials thus distracted, Baylor's thieving contingent rounded up hundreds of reservation horses, including those of local ranchers who had defied him.

After declaring that he would destroy the Indians if it "cost the life of every man in his command," Baylor began his retreat.[86] The very capable José María's fifty fighters shadowed Baylor's retreating party and saw his men lure a couple in their eighties to them. They were killed and scalped while the Indians watched. At this provocation, the chase was on, fifty Anadarko and Caddos exchanged fire with Baylor's fleeing vigilantes for eight miles until they reached the William Marlin ranch. Marlin escaped under a hail of gunfire while Baylor's men entered the house and began firing at the pursuers. José María's men were careful not to fire into the main ranch house because Marlin's wife and children were inside.[87] Mrs. Marlin later described the scene: Baylor's men were so badly frightened that some prayed, and one took up the boards to hide under the floor and the next day they buried their dead (seven) in the yard instead of the nearby cemetery.[88]

After stealing Marlin's horses, Baylor's vigilantes retreated to Rock Creek, where they formed minute companies and disbanded. Neighbors learned that it was Baylor's core of "desperadoes and horse thieves" that dominated the hundreds of hysterically misinformed settlers. A revelation of these motives caused many people to break away from the vigilante movement, especially since the dissolution of the reservations was well-known to be on the horizon.[89] On June 5, Baylor, incorrectly thought to be in command of a thousand vigilantes, ordered a feint at the Comanche reserve while his men rustled livestock from Anglo ranchers and distracted reservation Natives.[90] The governor of Texas, Hardin Runnels, continued to look the other way.

The Texas Trail of Tears

It was a rainy day on July 31, 1859, when the Caddos, Anadarkos, and others departed on their own "trail of tears," out of Texas to Oklahoma.[91] The 370 remaining Penateka Comanches and their remaining horses, federal troops, and Indian agents set out at the same time in a separate column. Altogether there were nearly 1,500 of the last Texas Natives and the Delawares and Shawnees that had married into them. The two columns met at the Red River on August 7, 1859, and spent the day crossing the river while looking over their shoulders for the threatened attack by Baylor's army.[92] Neighbors wrote: "I have this day crossed all the Indians out of the heathen land of Texas and am now out of the land of the Philistines."[93]

The Texas Indians and Neighbors were pleased with their new home on the Washita River: "This is in my judgment is a splendid country. The valleys are from one to five miles wide on alternate sides of the Washita. The soil, to judge from the heavy coat of grass and weeds, is very rich."[94] Neighbors resigned from his career as superintendent of Indian Affairs and was considering applying for agricultural instructor for his former charges. He asked for a leave to bring his family from San Antonino to the new reservation to "go to work in real earnest" for these people.[95]

On September 6, Neighbors descended into a very dangerous Texas; on September 13 he and his party made camp near Fort Belknap. The next morning Neighbors insisted on riding into the village alone. On arrival in the small town, Neighbors was confronted by one of Baylor's men, Patrick Murphy, angry because Neighbors had implicated him in horse thievery. Before Neighbors could respond, Murphy's brother-in-law, Edward Cornett, placed a double-barreled shotgun to Neighbor's back and pulled the trigger.[96] "Neighbors had incurred the vengeful animosity of many Texans by his zealous and uncompromising efforts to protect the Indians."[97]

The punishment inflicted by the Ford–Van Dorn expeditions and Baylor's violence put Bosque River settlers on alert for revenge raids. In February of 1860, there was a horrific raid in Erath County in which four hundred horses were stolen and seven people were killed. Also, two young women, the Lemley sisters, were kidnapped, raped, and abandoned, naked but alive. When questioned, the girls said the renegades spoke English and were not Plains Indians. Governor Houston responded to the outcry and sent Captain Middleton Tate Johnson to the Fort Cobb area to find evidence of those responsible. After two months of searching houses and examining horse herds, he found that Comanches had not left their camps on the upper Arkansas River. But for political reasons, Houston needed to punish somebody. Future Texas governor Lawrence Sullivan "Sul" Ross was among a detachment of Second Cavalry, a band of "self-sacrificing patriots," and Rangers who advanced into a remnant of Comancheria on the Pease River during December 1860 to carry out retribution.[98]

The Last of the Comanches

The December 19, 1860, massacre that followed took place on Mule Creek, a few miles upstream from the junction with the Pease River, and was soon spun into the collective memory of Texans. What was different about this short, brutal engagement was that truth was not among the survivors.[99] After the customary declaration that there was a trail leading to these people, the early morning attack overran a small Comanche hunting camp consisting of nine "grass tents" and about fifteen people, mostly women and children, just as they were packing to leave. Four women and three men were killed and some say scalped. The others mostly escaped because the Second Cavalry was reluctant to kill inoffensive women running away, and the Anglos suffered no casualties.[100] Around the campfires on the party's return to Texas, quite a different story would emerge.

The brief, one-sided attack on a small hunting camp morphed into a major Texas Ranger victory against five or six hundred fierce Comanche warriors.[101] One of the memorable, if fabricated, moments occurred when Sul Ross faced off with the redoubtable war chief Peta Nacona (he was not there), husband of Naduah (Cynthia Ann Parker) and father of Quanah Parker. According to the Rangers, Peta Nacona was escaping, riding double, when the rear rider, a girl, was shot, dragging the chief from the horse. Peta Nocona was on his feet instantly, shooting arrows at Ross, whose return fire broke the chief's right arm. Ross dismounted and continued shooting the Comanche until he began his "wild, weird death chant," at which time he allowed a Mexican servant to administer the "coup de grace."[102] The politically ambitious Ross created this story to pad his résumé to become a two-term governor of Texas (1887–91).[103]

Other aspects of the embarrassing event were invented to become part of Texas public memory; the capture of Cynthia Ann Parker and the narrow escape of Quanah Parker (who was not there either). Ross claimed it was his Rangers, not the Second Cavalry, who captured an old woman that turned out to be Cynthia Ann Parker.[104] One wonders how many narratives from the Texas frontier were created to serve agendas and to entertain children in later years, completely obscuring reality. Frontier apologist James T. DeShields shaped the narrative that turned myth into history: "So signal a victory that had never before been gained over the fierce and warlike

Comanches, and never since that fatal December day in 1860 have they made any military demonstrations at all commensurate with the fame of their proud campaigns in the past."[105]

The Civil War began in March 1861, bringing more violence to the upper Bosque River valley and Central Texas as the fighting men went to war. Already underway, the Civil War drought (1855–64), which was the worst in three hundred years, was centered on the Texas frontier and western Oklahoma, devastating the already reduced bison herds.[106] Thousands of men moved west to avoid conscription as many settlers were moving east to avoid the chaos and violence. Drought made sedentary life so difficult on the Texas frontier that many of these men joined the already numerous and organized livestock thieves.[107] These gangs, led by men like Patrick Murphy and Edward Cornet, operated through the Civil War, effectively disguised as Indians.[108] Adding to the violence, the Comanches, Kiowas, and Kickapoos turned to the frontier for supplies no longer available because of the collapsing hide trade.[109]

The Civil War drought, another persistent La Niña event, was especially lethal to the bison herds. Normally, bison take to the better grass in the river valleys, but this time those ecological niches were occupied by huge Comanche horse herds.[110] Unable to reach the best grasses the bison died in catastrophic numbers. The Comanches were still firmly locked into their faith that the bison could never become extinct, believing there were vast underground herds that overflowed each spring. Perhaps the bison had always produced more than their grassland would allow, and the Natives had served as a population check; this belief would explain why there were hardly any taboos against overhunting.[111] It's a rare case in history when a people correctly read the ecological warning signs and then actually take rational action to correct the situation.

For most Comanches, the market paradigm of bison robes for the American market seemed perfectly sustainable. Observers noted that Comanches "were badly custom-bound, caught up in magic or medicine" to the point that every Comanche reacted in virtually the same way in a given situation.[112] The Quahadis, the least affected by outsiders, led the way and nearly escaped the groove to become a pastoral society free of bison dependence. During the Civil War, bison collapse and constant diseases reduced the Comanches to five thousand people by the end of the war. To redefine themselves, the Comanches focused their raids on cattle, which they normally ignored or killed for revenge or meat, and less on horses.[113] In 1865 the Civil War and drought ended and the rains and grasses bounced back,[114] giving the Comanches one last chance for survival.

The return of the grasses reversed the bison collapse for a few years and continued the rapidly increasing unbranded longhorn numbers in Texas. The northern states had depleted their livestock to feed the United States military during the war, and the postwar burst in economic activity created a ravenous demand for Texas cattle. The Comanches became an important source of cattle for New Mexico and Colorado through both Anglos and Comanchero traders at several traditional rendezvous points.[115] History records a surge in Comanche cattle raids along the upper Bosque River at the close of the war, but a federal investigation concluded that only one of every twenty frontier deaths was caused by Indians. The Texas frontier was still a dangerous place because of the people of Texas.[116]

Six months after Appomattox, eleven northern Comanche chiefs met with government representatives on the Little Arkansas River in southern Kansas. The peace commissioners made the Comanches an offer they should not have refused. With the demands of the Texas legislators neutralized during Reconstruction, the United States was able to offer western Oklahoma and the Texas Panhandle as their reservation, where they could become full pastoralists. In return, they would be obliged to return all captives and stay within the reservation. Chief Eagle Drinking responded, "I am fond of the land I was born on. . . . The white man has enough land already."[117] The chief then refused to cede any land, yet the final Treaty of Little Arkansas did reaffirm forty thousand square miles of Texas for the Comanches. Two years later the Treaty of Medicine Lodge was interpreted differently by all sides. The Comanches should have taken the deal offered at the Little Arkansas.[118]

It became apparent to US peace commissioners that the Comanches were not ready to settle down, and during the winter of 1867–68, there were massive cattle raids in Texas. The herds were delivered to the Comancheros who traded tortillas, hard bread, other Pueblo/Mexican foods from the New Mexico Territory, Colt pistols, and whiskey from US cattle buyers, as well as coffee for cattle. The Comanches were competing with the still numerous cattle and horse rustling gangs from the days of the Baylor anarchy and even some diehard ex-Confederate cavalry.[119] The Comanches limited numbers were force-multiplied by Kiowas, Comancheros, some Cheyennes, and mixed-ethnic renegades.[120]

In 1868 the Senate authorized General William Tecumseh Sherman to administer Indian policy. General Sherman then authorized General Philip Sheridan to suppress the violence. Sheridan, who had just taken command of the Fifth Military District, had learned during the Civil War to strike the economic support as well as the enemy, and he ordered action against the Comancheros.[121]

The military pressure on the Comanchero trade began with the destruction of cattle herds of any New Mexicans on the western Llano Estacado, as well as the arrest of Anglo cattle herders in the same area. The Comanchero response was to disguise themselves as buffalo hunters, a protected group, to carry out their transactions in obscure Llano Estacado canyons, and they provided Comanche escorts for the cattle drives to the New Mexico Territory.[122]

The generals must have been frustrated when Ulysses S. Grant became president in 1869 and introduced his peace policy, the Quaker experiment, as it became known. This policy advocated less coercion and more Christian education through Protestant missionaries with a well-meaning but unimaginative Quaker, Lawrie Tatum, in charge of the Comanche-Kiowa agency.[123] This experiment ended badly when the policy of no military pursuits inside the reservations turned the agencies into demilitarized zones used by Comanches and Kiowas as a safe winter supply base after a season of raiding. For as long as he could, Tatum refused to comprehend that the people were better-fed in the off-reservation economy. He argued correctly that promised annuities were drained away by political corruption and indemnity payments for victims of the raids. The Quaker policy encouraged the ancient tradition of ransoming captives, mostly women, for one hundred dollars each in goods. The peace policy limited US military engagements with Indians unless they were caught in the act of raiding or transporting stolen cattle. Experienced Civil War cavalry under the regional commander, Colonel B. H. Grierson, were unable to react when Comanche war chief Tahbynaneekah, sometimes spelled Tabananica (Hears the Sunrise), challenged the soldiers to come out and fight.[124]

The Quaker experiment abruptly collapsed in 1871. A tour of frontier Texas by General Sherman barely escaped a mostly Kiowa war band that killed and mutilated seven teamsters near Fort Richardson on a road he had just traveled.[125] General Sherman was outraged by the unusually gruesome details; one man was tied upside down to a wagon wheel and roasted. Sherman began to ignore the Quaker peace experiment and sent Colonel Ranald S. Mackenzie into the Llano Estacado to resume the destruction of winter camps to disrupt the Comanchero cattle market.[126] Mackenzie and his Fourth Cavalry were sent illegally into Mexico, 1873–74, with the approval of President Grant, to attack suspected cattle raiders at Lipan and Kickapoo villages.[127]

In 1874, Colonel Mackenzie was back in northwestern Texas preparing for the final campaigns to push the Kiowas and Comanches to war and into the reservation.[128] Remembered as the Red River War, Mackenzie's first engagements intrigued and baffled his cavalry as the Comanches attacked in their wide V-formation to sweep around and fight horse to horse.[129] But Mackenzie's strategy was not, as among Texans, genocidal; he targeted Comanche resources—burning winter camps and supplies, capturing women and children (rather than killing them), and holding the women and children on reservations. Most effective of all was the arming and protection of buffalo hunters who had exterminated bison everywhere except south of the Arkansas River.[130] Bison were almost impossible for Kiowa and Comanche hunters to find, and, with reservation rations short, people on and off the reservation were starving.[131] The trauma was almost too much to bear. Something needed to be done to keep the Comanche world from dying.

Last Chance at Renewal: The Crisis Cult

Certain aspects of human nature lead to similar reactive social movements in the face of economic collapse. With limited cognitive ability to envision an adaptive pathway, people can become vulnerable to the irrational. Most commonly the reaction to severe stress is regression, the attempt to return to a more comfortable past. A universal response to this kind of stress opens a society to a charismatic cult leader who, empowered by a vision, speaks words with psychic resonance.[132] Such a leader rose to prominence on the Southern Plains between 1873 and 1874, an ecstatic shaman among the Quahadis Comanches. Isa-Tai (Coyote Droppings) experienced a prophetic vision in which he believed that he had ascended to the Great

Spirit and had been given the power to raise the dead and render bullets harmless.[133] The Comanches were in a mood to believe.

Joining a crisis cult requires stepping into a simpler world of belief that can offer peace and understanding, like a dream that refreshes the day-worn mind.[134] Isa Tai introduced the Comanche to the Plains Sun Dance in May 1874 for the expressed purpose of uniting and empowering almost all of the Comanches and other Southern Plains peoples. The simplified ceremony was to prepare the warriors for a war of extermination beginning with the center of bison hunter operations at William Bent's old trading post on the Canadian River, a thick-walled cluster of buildings known as Adobe Walls.[135] The ritualized journey to Adobe Walls included Isa-Tai-led songs, prayers, and dances, and on June 27, 1874, hundreds of Plains Indians were ready to attack at dawn.[136]

Isa-Tai, naked and painted yellow, sat on his horse while the war leaders, like Quanah Parker, charged the twenty-eight hunters over and over. The Anglos were armed with the latest long-range, large-bore buffalo rifles, which they fired from holes in the adobe. It was not the Plains Indian's way to dismount and maneuver for firing positions, a tactic they considered dishonorable. A Cheyenne who had just lost his son in one of these bold charges demanded that Isa-Tai ride in and recover his son's body if he were indeed bulletproof. Someone was knocked from his horse by a spent bullet fired from an impossible distance, and then the prophet's horse was killed. Isa-Tai claimed that his magic was ruined because someone killed a skunk the day before. Nobody was bulletproof; fifteen resistance fighters were killed, and Isa-Tai was ridiculed into obscurity.[137]

It was a crushing spiritual defeat for the Southern Plains peoples who had fully believed that the powers of the earth were behind them. Frustrated and angry, Comanches rode into Texas in a series of raids remembered as the Outbreak of 1874, which triggered a massive response. In late September, Colonel Mackenzie found hundreds of tipis along the Palo Duro and Blanca Cita Canyons in a natural trap. This was not a Texas-style genocidal raid. Hardly any Natives were killed, but tents and winter supplies were burned, and over a thousand horses were shot so that the Indians could not recapture them.[138] That fall and winter almost all of the Comanches came into the reservation, and, except for a couple of minor raids into Texas, 1875 ended the resistance.[139]

Quanah Parker and the New Road

Quanah Parker and the Quahadis avoided capture but understood that their situation was untenable; the soldiers had found their hiding places, and the bison were almost gone. Colonel Mackenzie sent out messengers to arrange surrender on June 2, 1875.[140] Mackenzie's humane treatment earned the respect of the Comanches, especially Quanah, who was, possibly because of his personal history, recognized as spokesman for all Comanches. In cooperation with Mackenzie, Parker rode out to negotiate with the last Comanche holdouts.[141] In July 1877, Quanah left Fort Sill with only three men, three women, and several mules loaded with supplies. With a letter from Colonel Mackenzie explaining their mission, they traveled under a white flag since there were plenty of Texans and buffalo hunters who would have been delighted to take Quanah's scalp.[142]

Quanah's party found the last sixty Comanche and Apache holdouts camped on the Pecos River in New Mexico. The leader of this tiny band was the German Comanche Herman Lehmann, the most reluctant man to surrender in the group.[143] It took four days to make it clear to the Indians that the Anglos would kill them if they found them. They cut their herd of horses and mules by three hundred so they could travel quickly at night to avoid Anglos. At the last minute, Herman Lehmann panicked, and for several months refused to surrender, working as a cowboy for Quanah. Eventually, he was discovered and returned to his family in Fredericksburg.[144] This time the Comanches began to adjust to farming and especially ranching.

Quanah was still in his twenties when he surrendered and began to help the Comanches and Kiowas find ways to improve their lives. Texas cattle were driven through their reservation on the way to market, fattened on the lush grassland. The Indians began to demand that good beef cattle be cut from the herd as tribute or passage fee to supplement meager reservation rations.[145] Quanah advocated leasing the grassland to cattlemen; a quasi-legal practice since it was technically government land. The *grass money*, paid to each Native in silver dimes, was legalized for another five years after Quanah and others traveled to Washington in 1885.[146] Quanah, with his functional English, negotiated with the cattlemen to provide Kiowa-Comanche employment for protection and herd management.[147] The goal of the reservation system was finally realized.

The last raids along the Bosque River during the winter of 1872 were less violent and more focused on livestock. In February, the district court was in session, and Stephenville was crowded with horses as people gathered for the court proceedings. During the night forty head of horses and mules were quietly taken, and no one saw who was responsible. Then in the spring of 1872, a cattle drive left the upper Bosque River headed for Colorado. A few miles out of town, Indians captured the herd and sent the cowboys back to Stephenville. A few days later another cattle drive started for the northwest, and this time a cowboy scouting ahead was killed and mutilated, and the herd was stolen.[148] Unidentified Indians were blamed for these raids.

The Comanches were indeed off the reservation in large numbers during the winter of 1872, including the same Anglo gangs from before the Civil War. It was their modus operandi to disguise themselves as generic Indians so convincingly that only after one was killed could he be identified as Anglo.[149] One of these gangs was confronted by an army patrol in December 1873 after rustling 120 horses near Fort Cobb. One gang member was captured; he offered details about how they made their disguises using bison hair. The patrol reported that they killed or captured a dozen "counterfeit white men in Indian disguise" but that the rustlers still numbered about a hundred men.[150] Cattle rustling and violence continued along the Bosque River, but no longer by Natives.[151]

There would be no law on the upper Bosque River for the next twenty years, and in response vigilante committees were organized. Sheriffs were routinely neutralized or murdered, leaving law enforcement to the "hooded men of justice" who dragged men from their houses to be hanged or summarily shot.[152] "Morality" was also enforced by vigilantes, houses were burned and settlers lectured on their behavior.[153] The state government placed Stephenville under martial law for several months sometime in the early 1870s, but no indictments were possible because "under the present reign of terror, no citizen ... will give evidence against the mob."[154] After the last sheriff was murdered, his deputy, John Gilbreath, was credited with ending vigilante law on the upper Bosque River in the mid-1880s.[155]

The Muddy Bosque River

Barbed wire was introduced on the Bosque River in the early 1880s, which brought an end to cattle companies that moved their herds from place to place, allowing the grasses to recover. Overgrazing followed until native grasses became all but extinct. Without ground cover, the land became disfigured with steep-sided arroyos as rainfall scoured the surface, muddying the Bosque River forever. By the early 1880s, the absence of grass fires opened the way for cedar (*Juniperus ashei*) to invade the lowlands, closing in on the prairies. Deep erosion lowered the water table, and the vegetation withered. Elk, antelope, and prairie chickens were gone, though a few alligators would keep arriving upstream for decades, and the last black bear was seen south of Stephenville in 1883.[156]

Cotton buyers controlled the economy, and cultivating the crop caused growers to exhaust the soil and degrade their lives until the boll weevil stopped production around 1950. Until the 1920s there were plenty of small farmers that survived by working less than forty acres. Dozens of tiny communities along the Bosque River were created, each with a store, a couple of churches, and a school, to hold the people together. A survey of the newspapers from this time describes impoverished farmers, surviving on the occasionally successful cotton crop, limited hog production, chickens, a milk cow, and garden produce. Reading the stories told in these newspapers shows a people tightly embedded in a church-centered, supportive network that many later looked back on as the good old days.

During the 1920s and 1930s, the small farmer's world fell apart. Ranching had been around since the first days of settlement, and in the early years, nobody believed any crop would grow besides turnips. Then ranchers slipped into the minority as Civil War veterans moved into the area as farmers in the 1880s. The economic depression that followed World War I, deepened during the Great Depression of the 1930s, favored beef producers, but not the farmers. These were droughty decades that made small family farms impossible, and it was during these years that the little communities faded away. Families began to experience real poverty as they were forced to sell everything, including the rock walls that once honeycombed the area, to rock crushers that soon covered the roads with the rubble. Farmers sold out to the ranchers and moved into the towns where they experienced culture shock at the loss of their tight communities. Ranches came to dominate the tired landscape, each ranch containing the remains of a dozen or more farmhouses. Once again, during the cycle of climate-caused replacement of one lifestyle with another, the small farmers (gatherers) gave way to the ranchers (bison hunters).

Notes

Chapter One

1. Susan Meriwether Harris, "The Western Cross Timbers; Scenario of the Past, Outcome for the Future" (master's thesis, Texas Christian University, May 2008), 14; and Douglas W. Hall, "Hydrologic Significance of Depositional Systems and Facies in Lower Cretaceous Sandstones, North-Central Texas," *Geological Circular* 76, no.1 (1976): 9–10.
2. Charles Frederick (geoarchaeologist), in discussion with the author, May 2019; and Brian J. Axsmith and Bonnie Fine Jacobs, "The Conifer Frenelopsis Ramosissima (Cheirolepidiaceae) in the Lower Cretaceous of Texas: Systematic, Biogeographical, and Paleoecological Implications," *International Journal of Plant Science* 166 (2005): 327.
3. Kim Ann Zimmermann, "Cenozoic Era: Facts about Climate, Animals, and Plants," *Live Science* 9 (2016).
4. Vaughn M. Bryant, Jr., and Richard G. Holloway, "A Late-Quaternary Paleoenvironmental Record of Texas: An Overview of the Pollen Evidence," in *Pollen Records of Late-Quaternary North American Sediments*, ed. Vaughn M. Bryant Jr. and Richard G. Holloway (Austin: American Association of Stratigraphic Palynologists Foundation, Austin, 1985), 49–50. The Pleistocene was composed of cooler full glacials and warmer interglacials, each lasting thousands of years. The vegetation is partly determined by the study of pollen (palynology) found in soil strata from each period.
5. Bryant and Holloway, *Pollen Records*, 49–50. Temperature and vegetation are confirmed by the presence or absence of animals like the giant beaver, long-nosed peccary, mastodon, mammoth, tapir, masked shrew, and bog lemming.
6. Donald A. Larson, et al., "Pollen Analysis of a Central Texas Bog," *American Midland Naturalist* 88 (October 1972): 358.
7. Mesic is a wetter environment, the opposite is xeric, dry.

8. Bryant and Holloway, *Pollen Records*, 52.
9. New analytical techniques have upset the usual paleoclimatic narrative with the revelation of how dramatic and abrupt these changes were at scales and rates not experienced in recorded history. Extinction events have been correlated around these warm-ups, especially the terminal Pleistocene at 12.8 thousand years ago. Alan Cooper et al., "Abrupt Warming Events Drove Late Pleistocene Holarctic Megafaunal Turnover," *Science* 349 (2015): 602; Darwin C. Hall et al., "Integrating Economic Analysis and the Science of Climate Instability," *Ecological Economies* 57 (2006): 443.
10. Bryant and Holloway, *Pollen Records*, 52–53.
11. Krista Clements Peppers, "Old Growth Forests in the Western Cross Timbers of Texas" (PhD diss., University of Arkansas, 2004), 126.
12. Peppers, "Old Growth Forests," 114, 122; Richard V. Francaviglia, *The Cast Iron Forest: A Natural and Cultural History of the North American Cross Timbers* (Austin: University of Texas Press, 2000), 39; James K. McPherson et al., "Competitive and Alleopathic Suppression of Understory by Oklahoma Oak Forests," *Bulletin of the Torrey Botanical Club* 99, no. 6 (1972): 273; and excavations have shown oak pollen to be often associated with charcoal in Pleistocene forests; Kim Wolfe, "Bur Oak (Quercus macrocarpa Michx.) in Riding Mountain National Park" (master's thesis, University of Manitoba, 2001), 43.
13. Jonathan Silvertown, *An Orchard Invisible: A Natural History of Seeds* (Chicago: University of Chicago Press, 2009): 78–83; and Walter D. Koenig and Johannes M. H. Knops, "The Mystery of Masting in Trees," *American Scientist* 93 (2005): 340–47.
14. Peter Wohlleben, *The Hidden Life of Trees: What They Feel, How They Communicate* (Vancouver: Greystone Books, 2015), 51; and Stephen Harrod Buhner, *The Lost Language of Plants: The Ecological Importance of Plant Medicines to Life on Earth* (White River Junction, VT: Chelsea Green Publishing, 2002), 162. I use the term *cloud* in the digital sense of unseen, stored activity and information.
15. Wohlleben, *The Hidden Life of Trees*, 49–55.
16. Wohlleben, *Hidden Life of Trees*, 49–55; and Francaviglia, *Cast Iron Forest*, 39.
17. Jennifer H. Carey, "*Quercus marilandica*": In: Fire Effects Information System," US Department of Agriculture, Forest Service, Rocky Mountain Research Station, Fire Sciences Laboratory https://www.fs.usda.gov/database/feis/plants/tree/quemar/all.html (May 24, 2019), 14, 46.
18. Wolfe, "Bur Oak," 43–52.

19. Jan Wrede, *Trees, Shrubs, and Vines of the Texas Hill Country* (College Station: Texas A&M University Press, 2015), 201; Darrell Sparks, "Adaptability of Pecan as a Species," *HortScience* 40, no. 5 (2005): 1182.
20. A problem with determining the time and exact location for the pecan's split from the other hickories is that the Carya pollen is all nearly identical. L. J. Grauke, "Geographic Patterns of Genetic Variation in Native Pecans," *Tree Genetics and Genomes* 7 (October 2011): 917; L. J. Grauke, "Hickory," in *A Guide to Nut Tree Culture in North America*, ed. Dennis Fulbright (Hamden, CT: Northern Nut Growers Association, 2003), 1–2.
21. Santiago Andres Catalano et al., "Molecular Phylogeny and Diversification History of Prosopis (*Fabaceae: Mimosoideae*)," *Biological Journal of the Linnean Society* 93, no. 3 (March 2008): 621–40; Jean-Pierre Simon, "Comparative Serology of a Disjunct Species Group: The Prosopis Juliflora-Prosopis Chilensis Complex," *Aliso: A Journal of Systematic Evolutionary Botany* 9, no. 3 (1979): 493–94.
22. Connie Barlow, *The Ghost of Evolution: Nonsensical Fruit, Missing Partners, and Other Ecological Anachronisms* (Berkeley: Perseus Books Group; Basic Books), 7, 156–57.
23. R. J. Ansley et al., "Mesquite Ecology," *Vernon, Texas: Agricultural Experiment Station* (March 18, 1997). https://texnat.tamu.edu/library/symposia/brush-sculptors . . . /mesquite-ecology/.
24. Prosopis pollen and charcoal were present in South and West Texas after the Pleistocene. The fire-friendly tree may have arrived in the Western Cross Timbers during the transition from woodland to scrub grasslands in West Texas between 14,000 and 10,000 years ago. Bryant and Holloway, *Pollen Records*, 58. Examinations of old growth Western Cross Timbers forests include mesquite; Krista Clements Peppers, "Old-Growth Forests" (PhD diss., University of Arkansas, 2004), 71; and early travelers described the mesquite as growing vertically rather than the lazy-limbed contemporaries that have grown up without fire. Lerena Friend, ed., *M. K. Kellogg's Texas Journal: 1872* (Austin: University of Texas Press, 1967).
25. In 1848 Ferdinand Roemer described prairies "scattered with mesquite trees"; cited in O. T. Hayward et al., *A Field Guide to the Grand Prairie of Texas: Land, History, Culture* (Waco, TX: Baylor Program for Regional Studies, 1992), 25.
26. Wrede, *Trees, Shrubs, and Vines*, 142, 201; M. L. Flint, ed., "Wild Blackberries," *Agriculture and Natural Resources* (Davis: University of California, 2019), 3.
27. Robert A. Vines, *Trees of Central Texas* (Austin: University of Texas Press, 1984), 310–11; and Wrede, *Trees, Shrubs, and Vines*, 188.

28. The name "Cast Iron Forest" was given by Washington Irving as he visited the Cross Timbers above the Red River; cited in Francaviglia, *The Cast Iron Forest*, 58–59.
29. Randolph B. Marcy, *Exploration of the Red River of Louisiana in the Year 1852* (US War Department, 1854), online: http://name.umdl.umich.edu/ABB2532.0001.001.
30. Elizabeth R. Gleim et al., "The Phenology of Ticks and the Effects of Long-Term Prescribed Burning on Tick Population Dynamics in Southwestern Georgia and Northwestern Florida," *PLoS ONE* 9, no. 11: e112174. https://doi.org/10.1371/journal.pone.0112174.
31. Logan A. West et al., "The Waco Mammoth National Monument May Represent a Diminished Watering-Hole Scenario Based on Preliminary Evidence of Post-Mortem Scavenging," *PALAIOS* 31 (2016): 592; Dan Flores, *American Serengeti: The Last Big Animals of the Great Plains* (Lawrence: University Press of Kansas 2016), 3; and in addition to the aforementioned animals, ancient terraces in the Bosque's sister river, Leon, reveal bones of two more grazing elephants, *Elephas boreus* and *Elephas imperator*, *Tapirus*, Ground sloth (*Mylodon harlan*), and the huge ground tortoise (*Testudo francisi*); in Raymond L. Lewand Jr., "The Geomorphic Evolution of the Leon River System," *Baylor Geological Studies* 17 (1969): 17.

Chapter Two

1. Now flooded and known as the Bering Strait, this area was exposed when Pleistocene glaciers held so much water that the sea level dropped two hundred feet. This landmass is called Beringia. Summer K. Praetorius, Jay R. Aler, et al., "Ice and Ocean Constraints on Early Human Migrations into North America along the Pacific Coast," *Proceedings of the National Academy of Sciences* 120, no.7 (February 6, 2023): e2208738120 https://doi.org/10.1073/pnas.2208738120.
2. Sarah Fowell and David Scholl, "The Bering Strait, Rapid Climate Change, and Land Bridge Paleoecology, Final report of the JOI/USSSP/IARC Workshop, June 20–22, 2005; Plio-Pleistocene, *American Museum of Natural History*, https://research.amnh.org/palentology; Scott Armstrong Elias, "First Americans Lived on Bering Land Bridge for Thousands of Years," *The Conversation, Scientific American* (March 4, 2014).
3. There are several American sites as old or older than 13,000 years ago, like Monte Verde, Gault, Friedkin, Paisley Caves, and Cooper's Ferry; Loren G.

Davis and David B. Madsen, "The Coastal Migration Theory: Formulation and Testable Hypotheses," *Quaternary Science Reviews* 249 (2020): 106605.

4. Jennifer Raff, *Origin: A Genetic History of the Americas* (New York: Hachette Book Group, 2022), 191; *plant fossils* refer to the glassy, fossilized part of a plant's structure; the insect remains were fossil beetle shells that have such narrow temperature requirements that they serve as buried thermometers; Scott A. Elias, "Late Pleistocene Climates of Beringia, Based on Analysis of Fossil Beetles," *Quaternary Research* 53, no. 2 (January 2017): 229–32; Lauriane Bourgeon and Ariane Burke, "Horse Exploitation by Beringian Hunters during the Last Glacial Maximum," *Quaternary Science Reviews* 269 (October 2021): 107140.

5. Some have reckoned that the "Bering Standstill" lasted from 5,000 to 8,000; Eske Willerslew and David J. Meltzer, "Peopling of the Americas as Inferred from Ancient Genomics," *Nature* 594 (June 2021): 356–61; Jim Scott, "INSTAAR-led Study Says Bering Land Bridge Area Likely a Long-Term Refuge for Early Americans," *Institute of Arctic and Alpine Research* (February 27, 2014): 4–5; Maanasa Raghavan, Matthias Steinrucken et al., "Genomic Evidence for the Pleistocene and Recent Population History of Native Americans," *Science* 349, no. 6250 (August 21, 2015): 1–2.; Raff, *Origin*, 191–92; Elias, "First Americans," 5.

6. Michael Wilson, "Geology in the Ice-Free Corridor," *A Journey to a New World*, last updated 2005, Simon Fraser University, www.sfu.museum/journey/an-en/postsecondaire-postsecondary/geologie_ifr; Lauren Milideo, "Fieldwork Revises Ice-Free Corridor Hypothesis of Human Migration," *Earth: The Science Behind the Headlines*, last updated April 13, 2014, www.earthmagazine.org/article/fieldwork-revises-ice-free-corridor-hypothesis-human-migration/#:~:text=The%20latest%20research%2C%20Swisher%20says,of%20an%20ice%2Dfree%20corridor.

7. Raff, *Origin*, 191.

8. David Meltzer, ed., Jorie Clark, Anders E. Carlson, and Alberto V. Reves, "The Age of the Opening of the Ice-Free Corridor and Implications for the Peopling of the Americas," *Proceedings of the National Academy of Sciences* 119, no. 14 (March 21, 2022): e2118558119.

9. Raff, *Origin*, 77–88; Jerome E. Dobson, Giorgio Spada, and Gaia Galassi, "The Bering Transitory Archipelago: Stepping Stones for the First Americans," *Geoscience* 353, no. 1 (2021): 55–65; David Bustos and Thomas M. Urban, "Reply to 'Evidence for Humans at White Sands National Park during the Last Glacial Maximum Could Actually be for Clovis People—13,000 Years Ago' by C. Vance Hayes, Jr.," *PaleoAmerica: A Journal of Early Migration and Dispersal* 8, no. 2 (2022); Meadowcroft, an archaeological site, is well-known.

10. Matthew S. Taylor, "The Midland Calvarium and the Early Human Habitation of the Americas," *Bulletin of the Texas Archeological Society* 86 (2015): 193, 205; Michael R. Waters, "Early Exploration and Settlement of North America during the Late Pleistocene," *SAA Archaeological Record* (May 2019): 35–36.
11. This was the Bølling-Allerød warm period; Michael J. O'Brien, "Setting the Stage: The Late Pleistocene Colonization of North America," *Quaternary* (2019): 2,1, doi:10.3390/quat2010001; pollen has been the gold standard for reconstructing past environments, but it can be blown hundreds of miles or just drop from a plant without registering in nearby sediment. There are beetles that signal temperature and moisture. Certain fauna, like the cotton rat, *Sigmodon hispidus*, have a narrow tolerance for environmental conditions. Phytoliths are formed when silica in solution surrounds a plant's cells when encased in sediment and hardens to preserve the exact plant identification in glass-like silica for millions of years. Vaughn M. Bryant Jr. and Richard G. Holloway, "A Late-Quaternary Paleoenvironmental Record of Texas: An Overview of the Pollen Evidence," in *Pollen Records of Late-Quaternary North American Sediments*, ed. Vaughn M. Bryant and Richard G. Holloway (Austin: The American Association of Stratigraphic Palynologists Foundation, 1985), 53–54.; Michael B. Collins and C. Britt Bousman, "Cultural Implications of Late Quaternary Environmental Change in Northeastern Texas," *CRHR Research Reports* 1, no. 6 (2015): 73; Frank C. Leonhardy, "Late Pleistocene Research at Domebo: A Summary and Interpretation," in *Domebo: A Paleo-Indian Mammoth Kill in the Prairie-Plains*, ed. Frank C. Leonhardy (Lawton, OK: Contributions of the Museum of the Great Plains, 1966): 52. The 24,000 to 14,000 BP climate range is from Corrine Wong and Jay Banner, "Holocene Climate Variability in Texas, USA: An Integration of Existing Paleoclimate Data and Modeling with a New, High-Resolution Speleothem Record," *Quaternary Science Reviews* 127 (2015): doq: 10.1016/j.quascirev.2015.06.023.
12. Frederik V. Seerholm, Daniel J. Werndly et al., "Rapid Range and Megafaunal Extinctions Associated with Late Pleistocene Climate Change," *Nature Communications* (2022): 1–10.
13. Joshua John Porter, "Reconstructing Bison and Mammoth Migration during the Late Pleistocene and Early Holocene of Central Texas Using Strontium Isotopes," (master's thesis, University of Arkansas, May 2022), 12; Jon Baskin, *The Pleistocene Fauna of South Texas* (Kingsville, TX: A&M University, 2004), users.tamuk.edu/kfjabo2/SOTXFAUN.htm; Whit Bronaugh, "The Trees That Miss the Mammoths," *American Forests* (2015), www.anericanforests.org/magazine/article/trees-that-miss-the-mammoths/.

14. O'Brien, "Setting the Stage," 5; Blake De Pastino, "16,000-Year-Old Tools Discovered in Texas, Among the Oldest Found in the West," *Western Digs*, last updated July 18, 2016, http://westerndigs.org/16000-year-old-tools-in-texas-among-oldest-yet-found-in-the-west/15; Irina Y. Ponkratova, Loren J. Davis, et al., "Technological Similarities between 13ka Stemmed Points from Ushki V, Kamchatka, Russian Far East, and the Earliest Stemmed Points in North America," chapter 4 in *Maritime Prehistory of Northeast Asia*, ed. Jim Cassidy, Irina Ponkratova, and Ben Fitzhugh (Singapore: Springer, 2022), 233–61; T. A. Surovell, S. A. Allaun et al., "Late Date of Human Arrival to North: Continental Scale Differences in Stratigraphic Integrity of Pre-13,000 BP Archeological Sites," *PLoS ONE* 17, no. 4 (2022); Katy Dycus, "After Cooper's Ferry, Rethinking How the Americas Were Peopled," *Mammoth Trumpet: Center for the Study of the First Americans* 38, no. 2 (April 2023): 8–20.
15. D. Clark Wernecke, past director of the Gault School of Archaeological Research, email communication (July 2017): there were no big bones in the pre-Clovis level at Gault; I don't know about bones at other pre-Clovis locations.
16. The point technology represented on Buttermilk Creek has been interpreted as stages of development from the Western Stemmed look-alikes, to lanceolate and then to Clovis; Jordan Pratt, Ted Goebel, et al., "A Circum-Pacific Perspective on the Origin of Stemmed Points in North America," *PaleoAmerica* 6, no. 1 (2020); Michael R. Waters, Joshua L. Keene, et al., "Pre-Clovis Projectile Points at the Debra L. Friedkin Site, Texas Implications for the Late Pleistocene Peopling of the Americas," *Science Advances* 4, no. 10 (October 24, 2018): 4; John D. Speth, Kori Newlander, et al., "Early Paleoindian Big-Game Hunting in North America: Provisioning or Politics?" *Quaternary International* (2010): 22; Michael R. Waters, Thomas W. Stafford, et al., "The Age of Clovis—13,050—12,750 cal yr B.P.," *Science Advances* 6, no. 43 (October 21, 2020): eaaz0455; Wernecke, personal communication (August 2019), holds that the Clovis points are tools used by different peoples of this era with no cultural attachments.
17. Briggs Buchanan, Brian Andrews, et al., "An Assessment of Stone Weapon Tip Standardization during The Clovis-Folsom Transition in the Western United States," *American Antiquity* 83, no. 4 (2018): 5; Speth, "Early Paleoindian Big-Game Hunter," 5–7.
18. A red hydrated iron oxide was valued all over the prehistorical world as symbol-laden body paint, especially regarding the dead. Clovis points in Bosque River collections remind us that sophisticated people frequented the area.

19. Matt Ridley, *The Rational Optimist: How Prosperity Evolves* (New York: HarperCollins, 2010), 54–58; R. E. Whallon, "Social Networks and Information: Non-Utilitarian Mobility among Hunter-Gatherers," *Journal of Anthropological Archaeology* 25, no. 2 (2006): 259.
20. Susan Dial and Albert Redder, "A Paleoindian Grave: The Horn Shelter," The University of Texas at Austin, last updated December 2010. https://www.texasbeyondhistory.net/horn/burials.html.
21. Atlatl remains have not yet been found among Clovis or earlier sediments, but the use of the atlatl is inferred by the impact fractures in the chert point, that could not have been made by a hand-thrown spear; Joseph Castro, "First Americans Used Spear-Throwers to Hunt Large Animals," *Live Science*, last updated January 28, 2015, www.livescience.com/49603-paleoindian-spear-thrower-evidence.html; Metin I. Eren, David J. Meltzer, et al., "On the Efficacy of Clovis Fluted Points for Hunting Proboscideans," *Journal of Archeological Science: Reports 39* (October 2021): 103166.
22. Speth, "Early Big-Game Hunters,"14, 18.
23. My speculation.
24. It's my understanding that the mild Pleistocene lacked flooding events and people camped much closer to streams than later peoples, and that the fires weren't curbed with rocks as was the custom later. The rockless campfires came from conversations with Charles Frederick.
25. Nina Munteanu, "The Power of Myth in Storytelling: The Alien Next Door," 2017, https://ninamunteanu.me/2017/03/28/the-power-of-myth-in-storytelling-2/.
26. Inferred from D. Clark Wernecke and Michael B. Collins, "Patterns and Process: Some Thoughts on the Incised Stones from the Gault Site, Central Texas, United States," *Pleistocene Art of the World: Symposium*, September 2010, 678.
27. Wernecke and Collins, "Patterns and Process," 670.
28. Ridley, *Rational Optimist*, 52–54; and Ashley K. Lemke, D. Clark Wernecke, et al., "Early Art in North America: From the Clovis and Later Paleoindian Incised Artifacts from the Gault Site, Texas (41BL323)," *American Antiquity* 80, no. 1 (2015): 128.
29. Inferred from Wernecke and Collins, "Patterns and Process," 671, 675.
30. Clark Spencer Larsen, *Skeletons in Our Closet: Revealing Our Past through Bioarchaeology* (Princeton, NJ: Princeton University Press, 2000), 98, 112–13.
31. Glenn Hodges, "First Americans," *National Geographic Magazine*, January 2015, 3–4; and Randall Haas, James Watson, et al., "Female Hunters of the Early Americas," *Science Advances* 6, no. 45 (November 4, 2020).
32. Larsen, *Skeletons*, 98, 112–13.

33. Larsen, *Skeletons*, 52, 112–13; James C. Chatters, Douglas J. Kennett, et al., "Late Pleistocene Human Skeleton and mtDNA from Mexico Links Paleo-americans and Modern Native Americans," *Science* 344, no. 6185 (2014): 750–54. The visage assertion is inferred, but it seems reasonable. A smaller mouth means crowded wisdom teeth. Nearly every time I go to the dentist I ask if they know how many people are currently born without wisdom teeth. I haven't checked the studies they used, but they assure me that the number increases all the time.
34. Minxia Lu, Liang Chen, et al., "A Brief History of Wheat Utilization in China," *Frontiers of Agricultural Science and Engineering* 6, no. 3 (2019): 289; Frederik V. Seersholm, Daniel J. Werndly, et al., "Rapid Range Shifts and Megafaunal Extinctions Associated with Late Pleistocene Climate Change," *Nature Communications* 11 (2020): 1–7.
35. "Microscopic Diamonds Suggest Cosmic Impact Responsible for Major Period of Climate Change," *Geology Times* 1 (September 15, 2014); Christopher R. Moore, Allen West, et al., "Widespread Platinum Anomaly Documented at the Younger Dryas Onset in North American Sedimentary Sequences," *Scientific Reports* 7 (2017): 1; James Lawrence Powell, "Premature Rejection in Science: The Case of the Younger-Dryas Impact," *Science Progress* 105, no. 1 (January 5, 2022).
36. The Younger Dryas dates, 12,800 to 11,700 years ago, are averaged from the many studies. Lucinda McWeeney, "Revising the Paleoindian Environmental Picture in Northeastern North America," in *Foragers of the Terminal Pleistocene in North America*, ed. Renee B. Walker and Boyce N. Driskell (Lincoln: University of Nebraska Press), 165; Paul Voosen, "Impact Crater under Greenland's Ice Is Surprisingly Ancient," *Science* 375, no. 6585 (March 11, 2022): 1076–77; Martin B. Sweatman, "The Younger Dryas Impact Hypothesis: Review of the Impact Evidence," *Earth Science Reviews* 218 (July 2021): 103677; Powell, "Premature Rejection"; David Tovar Rodriguez, "A Possible Late Pleistocene Impact Crater in Central North America and Its Relation to the Younger Dryas Stadial" (master's thesis, University of Minnesota, 2020). 40–42.
37. Nan Sun, Allan D. Brandon, et al., "Geochemical Evidence for Volcanic Signatures in Sediments of the Younger Dryas Event," *Geochimica et Cosmochimica Acta* 312 (August 2021).
38. P. M. Abbott, U. Niemeier, et al., "Volcanic Climate Forcing Preceding the Inception of the Younger Dryas: Implications for Tracing the Laacher See Eruption," *Quaternary Science Reviews* 274 (December 1, 2021): 107260; N. Sun, A. D. Brandon, et al., "Volcanic Origin for Younger Dryas Geochemical Anomalies ca. 12,900 B.P.," *Science Advances* 6 (2020): 1–9;

S. L. Norris, D. Garcia-Castellanos, et al., "Catastrophic Drainage from the Northwestern Outlet of Glacial Lake Agassiz during the Younger Dryas," *Geophysical Research Letters* (2020): 1–2; Mark Thiemens, ed., Hai Cheng, Haiwei Zhang, et al., "Timing and Structure of the Younger Dryas Event and Its Underlying Climate Dynamics," *Proceedings of the National Academy of Sciences* 117, no. 38 (September 20, 2022): 1–10; David J. Leydet et al., "Opening of Glacial Lake Agassiz's Eastern Outlets by the Start of the Younger Dryas Cold Period," *Geological Society of America* (2018).

39. Seersholm and Werndly, "Rapid Range Shifts," 1–10.
40. Seersholm and Werndly, "Rapid Range Shifts"; Briggs Buchanan, J. David Kilby, et al., "Bayesian Modeling of the Clovis and Folsom Radiocarbon Records Indicates a 200-Year Multigenerational Transition," *American Antiquity* (2022): 1–14; Carlos E. Cordova and William C. Johnson, "An 18ka to Present Pollen- and Phytolith-Based Vegetation Reconstruction from Hall's Cave, South-Central Texas, USA," *Quaternary Research* 92, no. 2 (May 7, 2019); "Clovis Reconsidered: The Gault Site," www.texasbeyondhistory.net/gault/clovis.html; Todd A. Surovell, Joshua R. Boyd, et al., "On the Dating of the Folsom Complex and Its Correlation with the Younger Dryas, the End of Clovis, and Megafaunal Extinction," *PaleoAmerica* 2, no. 2 (2016): 81–89; Seerholm and Werndly, "Rapid Range Shifts," 2770. https://doi.org/10.1038/s41467-020-16502-3.
41. Lemke and Wernecke, "Early Art"; Collins, "Patterns and Process"; Waldo R. Wedel, "Culture Sequence: The Central Great Plains," *Plains Anthropologist* 17, no. 57 (1972): 291–354.
42. Buchanan, "Bayesian Modeling."
43. Briggs Buchanan, Brian Andrews, et al., "Settling into the Country: Comparison of Clovis and Folsom Lithic Networks in Western North America Shows Increasing Redundancy of Toolstone Use," *Journal of Anthropological Archaeology* 53 (2019): 32–42; Anthony T. Boldurian and Susanne M. Hubinsky, "Preforms in Folsom Lithic Technology: A View from Blackwater Draw, New Mexico," *Plains Anthropologist* 39, no. 150 (November 1994): 445–64; Andrew J. Richard, "Clovis and Folsom Functionality Comparison," (master's thesis, University of Arizona, 2015), 59.
44. I taught an advanced Texas History class for years in which students were given credit for constructing and throwing with atlatls. The turkey-fletched shafts were made from river cane, and the foreshafts were oak. The range limit seemed to be 110 yards, but I've heard claims of longer throws.
45. Metin I. Eren, Michelle R. Bebber, et al., "Plains Paleoindian Projectile Point Penetration Potential," *Journal of Anthropological Research* 78, no. 1 (2022); Stanley A. Ahler, "Why Flute? Folsom Point Design and Adaption," *Journal of Archaeological Science* 27 (2000): 804–11.

46. Buchanan, "Settling into the Country"; Brendon Patrick Asher, "From the Continental Divide to the Plains-Woodland Border: Clovis and Folsom/Midland Land Use and Lithic Production" (PhD diss., University of Kansas, 2015), 173.
47. Among the returning flora and fauna to Central Texas were two not before detected in archaeological sites, the raccoon (*Procyon lotor*) and the redbud tree (*Cercis*); Seerholm and Werndly, "Rapid Range Shifts."

Chapter Three

1. C. Britt Bousman, Michael B. Collins, et al., "The Paleoindian-Archaic Transition in North America: New Evidence from Texas," *Antiquity* (January 15, 2002): 980–90; Thornton R. Raskevitz, "Phytolith Analysis as a Paleoecological Proxy When Examining Bison Anatomical and Behavior Changes in the Great Plains" (master's thesis, Oklahoma State University, 2020).
2. I. S. O. Matero, L. J. Gregoire, et al., "The 8.2 ka Cooling Event Caused by Laurentide Ice Saddle Collapse," *Earth and Planetary Science Letters* 473 (September 1, 2017): 205–14.
3. Alston V. Thoms, "Ancient Savannah Roots of the Carbohydrate Revolution in South-Central North America," *Plains Anthropologist* 53 (February 8, 2008): 121–36, no. 205; Alston V. Thoms, "Rocks of Ages: Propagation of Hot-Rock Cookery in Western North America," *Journal of Archeological Science* 36 (2009): 573–91.
4. Alston V. Thoms, "Ethnographies and Actualistic Cooking Experiments: Ethnoarchaeological Pathways toward Understanding Earth-Oven Variability in Archaeological Records," *Ethnoarchaeology* 10, no. 2 (2018): 76–98.
5. Ellen Sue Turner and Thomas R. Hester, *A Field Guide to Stone Artifacts of Texas Indians* (Houston: Gulf Publishing, 1985); projectilepoints.net; observations by Dan Young of Bosque River collections, 1965–Present.
6. Turner and Hester, *Field Guide*, 1985; Young, observations; and projectile points.net.
7. Harry J. Shafer and Steve A. Tomka, "Early Middle Archaic Lithic Assemblages," in *An Early Middle Archaic Site along Cordova Creek in Comal County, Texas*, ed. Richard B. Mahoney, Harry J. Shafer, et al., prepared for Texas Department of Transportation Environmental Affairs Studies Program, Report number 49, and Center for Archaeological Research, *The University of Texas at San Antonio Archaeological Survey Report*, no. 332 (2003).

8. Darrell Kaufman, Nicholas McKay, et al., "Holocene Global Mean Surface Temperature, a Multi-Method Reconstruction Approach," *Scientific Data*, 7, no. 201 (June 30 2020); David J. Meltzer, "Human Responses to Middle Holocene (Altithermal) Climates on the North American Great Plains," *Quaternary Research* 52 (1999): 404–16; M. Gaetani, G. Messori, et al., "Mid Holocene Climate at Mid-Latitudes: Modeling the Impact of the Green Sahara," *EGU General Assembly Abstracts* (May 2022): EGU22–2031; Ming Zhang, Yonggang Liu, et al., "AMOC and Climate Responses to Dust Reduction and Greening of the Sahara during the Mid-Holocene," *Journal of Climate* 34, no. 12 (2021): 4893–912.
9. Meltzer, *Human Responses*; Kaufman et al., *Holocene Global*; G. W. Tomanek and G. K. Hullet, "Effects of Historical Droughts on Grassland Vegetation in the Central Great Plains," in *Pleistocene and Recent Environments of the Central Great Plains*, ed. Wakefield Dort Jr. (Lawrence: University Press of Kansas, 1970), 205–9; and Eileen Johnson, Stance Hurst, et al., "Late Quaternary Stratigraphy and Geochronology of the Spring Creek Drainage along the Southern High Plains Eastern Escarpment, Northwest Texas," *Quaternary* 4, no. 19 (June 22, 2021): 1–37.
10. Rebecca Taormina, Lee Nordt, et al., "Late Quaternary Alluvial History of the Brazos River in Central Texas," *ScienceDirect* 631 (September 10, 2022): 34–46; Christopher L. Odezulu, Travis Swanson, et al., "Holocene Progradation and Retrogradation of the Central Texas Coast Regulated by Alongshore and Cross-Shore Sediment Flux Variability," *Depositional Record* 7, no. 1 (November 20, 2020): 77–92; Johnson, *Late Quaternary Stratigraphy*.
11. Lucas C. Majure, Paul Puente, et al., "Phylogeny of Opuntia s.s. (Cactaceae): Clade Delineation, Geographic Origins, and Reticulate Evolution," *American Journal of Botany* 99, no. 5 (April 2012): 847–64; Matt Warnock Turner, *Remarkable Plants of Texas: Uncommon Accounts of Our Common Natives* (Austin: University of Texas Press, 2009), 251–54.
12. Phil Dering, "Daily Bread and Healing Balm: A Deep History of Native Plant Use in the Trans-Pecos of Texas," *The Sabal* 23, no. 1 (January 2006); Turner, *Remarkable Plants*, 248–56.
13. Delena Tull, *Edible and Useful Plants of Texas and the Southwest* (Austin: University of Texas Press, 1987), 51–52.
14. Kelly Kindsher, *An Ethnobotanical Guide: Edible Wild Plants of the Prairie* (Lawrence: University Press of Kansas, 1987), 221; Willis Harvey Bell and Edward Franklin Castetter, *The Utilization of Yucca, Sotol, and Beargrass by the Aborigines in the American Southwest* (Albuquerque: University of New Mexico, Ethnobotanical Studies in the Southwest, 1941): 3, 5.

15. Steve Black and Vaughn M. Bryant Jr., "Hinds Cave: A Perishable Scientific Treasure," *Texas Beyond History*, University of Texas at Austin, 2005; Karl J. Reinhard et. al., "Hunter Gatherers Use of Small Animal Food Resources: Coprolite Evidence," *International Journal of Osteoarchaeology* 17 (2007): 421.
16. Karl J. Reinhard and Vaughn M. Bryant Jr., "Coprolite Analysis: A Biological Perspective on Archeology," *Papers in Natural Resources* (1992): 253–54.
17. This is a reference to the Buckeye Knoll Site in South-Central Texas; J. Colette Berbesque and Kara C. Hoover, "Frequency and Developmental Timing of Linear Enamel Hypoplasia Defects in Early Archaic Texan Hunter-Gatherers," *PeerJ6* (2018): e4367; Larsen, *Skeletons in Our Closet*, 16.
18. Stephanie Pappas, "Ancient Poop Gives Clues to Modern Diabetes Epidemic," *Live Science* (July 24, 2012); Karl J. Reinhard, "Understanding the Pathological Relationship between Ancient Diet and Modern Diabetes through Coprolite Analysis: A Case Example from Antelope Cave, Mojave County, Arizona," *Papers in Natural Resources*, Paper 321 (2012).
19. Pappas, "Ancient Poop."
20. Karl J. Reinhard and Vaughn M. Bryant Jr., "Pathology and the Future of Coprolite Studies in Bioarchaeology," *Papers in Natural Resources*, Paper 43 (2008): 212–13; Dennis Danielson and Karl Reinhard, "Human Dental Microwear Caused by Calcium Oxalate Phytoliths in Prehistoric Diet of the Lower Pecos Region, Texas," *American Journal of Anthropology* 107 (1998): 297–98.
21. Anna Starzynska, Piotr Wychowanski, et al., "Association between Maternal Periodontitis and Development of Systematic Diseases in Offspring," *International Journal of Molecular Sciences* 23, no. 5 (2022): 2473; Andrea Horvath Marques, Thomas G. O'Conner, et al., "The Influence of Maternal Prenatal and Early Childhood Nutrition and Maternal Prenatal Stress on Offspring Immune System Development and Neurodevelopmental Disorders," *Frontiers in Neuroscience* 7 (July 31, 2013); Berbesque, "Frequency and Development"; Leland C. Bement, *Hunter-Gatherer Mortuary Practices during the Central Texas Archaic* (Austin: University of Texas Press, 1994), 95.
22. Berbesque, "Frequency and Development."
23. Pappas, "Ancient Poop."

Chapter Four

1. Bryan N. Shuman, "Millennial Variations and a Mid-Holocene Step Change in Northern Mid-Latitude Moisture Gradients," *Research Square* (February 19, 2021): 7–8; John C. Lohse, Don G. Wyckoff, and Marjorie Duncan, "The

Calf Creek Horizon: Mid-Holocene Adaptions in North America," in *The Calf Creek Horizon: A Mid-Holocene Hunter-Gather Adaption in the Central and Southern Plains of North America*, ed. John C. Lohse et al., chapter 1 (College Station: Texas A&M University Press, 2021), 4–8.
2. Lohse, Wyckoff, and Duncan, "Calf Creek Horizon," 4; and Kristen Carlson and Leland Bement, "Bison across the Holocene: What Did Calf Creek Foragers Hunt?" in *Calf Creek: A Mid-Holocene Adaption*, ed. Lohse et al., chapter 2, 16–17.
3. Lars Zver, Borut Toskan, and Elena Buzan, "Phylogeny of Late Pleistocene and Holocene *Bison* Species in Europe and North America," *Quaternary International* 592 (2021): 30; and Duane Froese, Mathias Stiller, et. al., "Fossil and Genomic Evidence Constrains the Timing of Bison Arrival in North America," *Proceedings of the National Academy of Sciences* 114, no. 13 (March 28, 2017): 3457–62.
4. Lohse, Wyckoff, and Duncan, "Calf Creek Horizon," 3–15.
5. Elton R. Prewitt, "Bell: A Calf Creek Series Dart Point Type in Texas," in Lohse, Wyckoff, and Duncan, "Calf Creek Horizon," 147.
6. Sergio Ayala, "Technology and Typology of the Calf Creek Horizon," in Lohse, Wyckoff, and Duncan, "Calf Creek Horizon."
7. Ayala, "Technology and Typology," 100–101.
8. J. Michael Quigg, Paul M. Matchen, et. al., "Eligibility Assessment of the Slippery Slope Site (41MS69) in TxDot Right-of-Way in Mason County, Texas," Index of Texas Archaeology: Open Access Gray Literature from the Lone Star State, vol. 2015, 17–19, 30.
9. Leland C. Bement, *Hunter-Gatherer Mortuary Practices during the Central Texas Archaic* (Austin: University of Texas Press, 1994); Karl W. Kibler and Tim Gibbs, "Archeological Survey of 61 Acres along the Bosque River, Waco, McLennan County, Texas," Ross C. Fields, Principal Investigator, Prewitt and Associates Inc., Technical Reports, Number 69 (July 2004), 4; and Karl Kibler and Gemma Mehalchick, "Hunter-Gatherer Resource Acquisition and Use in the Lower Bosque River Basin during the Late Archaic," *Bulletin of the Texas Archeological Society* 81 (2010): 113.
10. Kibler and Mehalchick, "Hunter-Gatherer Resource Acquisition," 470; and Jurgen Bader, Johann Jungclaus, et al., "Global Temperature Modes Shed Light on the Holocene Temperature Conundrum," *Nature Communications* 11 (2020): 2.
11. Matt Warnock Turner, *Remarkable Plants of Texas: Uncommon Accounts of Our Common Natives* (Austin: University of Texas Press, 2009), 67–69; and R. J. Ansley, J. A. Huddle, et al., "Mesquite Ecology," *Texas Natural Resources Server*, 1–2.

12. Turner, *Remarkable Plants of Texas*; and "Mesquite," *Texas Beyond History*, University of Texas at Austin: Texas Archeological Research Laboratory, since 2001.
13. Walter D. Koenig and Johannes M. H. Knops, "The Mystery of Masting in Trees," *American Scientist* 93 (2005): 347; and Lohse, Wyckoff, and Duncan, "Calf Creek Horizon," 10.
14. Michael David Sabrin, "Characterization of Acorn Meal" (master's thesis, University of Georgia, 2009), 8–9; Delena Tull, *Edible and Useful Plants of Texas and the Southwest* (Austin: University of Texas Press, 1987), 102; and Turner, *Remarkable Plants of Texas*, 78.
15. Kibler and Gibbs, "Archeological Survey," 4; Bement, *Mortuary Practices*, 14; and projectilepoints.net.
16. Bement, *Mortuary Practices*, 90, 115–17.
17. Bryan N. Shuman, "Millennial Variations and a Mid-Holocene Step Change in Northern Mid-Latitude Moisture Gradients," *Research Square* (2021): 1–25; and Raymond S. Bradley and Jostein Bakke, "Is There Evidence for a 4.2 ka BP Event in the Northern North Atlantic Region?" *Climates of the Past* 15 (2019): 1665–76.

Chapter Five

1. Chijun Sun, Timothy Shanahan, et al., "Spring Warming Drives Past and Future Shifts in Great Plains Storm Intensity," *Nature Geoscience* (November 29, 2021): 5; Christopher B. Skinner, Juan M. Lora, et al., "Atmospheric River Changes Shaped Mid-Latitude Hydroclimate since the Mid-Holocene," *Earth and Planetary Science Letters* 541 (July 1, 2020): n.p.; and Heinz Wanner, Jurg Beer, et al., "Mid- to Late Holocene Climate Change: An Overview," *Quaternary Science Reviews* 27, nos. 19–20 (October 2008): 1791–828.
2. Gemma Mehalchick and Karl W. Kibler, "Hunters and Gatherers of the North Bosque River Valley: Excavations at the Baylor, Britton, McMillan, and Higginbotham Sites, Waco Lake, McMillan County, Texas," Prewitt and Associates Inc., Cultural Resources Services, Austin, Texas (July 2008), 9, 14, 19.
3. Walter D. Koenig and Johannes M. H. Knops, "The Mystery of Masting in Trees," *American Scientist* 93 (2005): 344–45.
4. Matt Warnock Turner, *Remarkable Plants of Texas: Uncommon Accounts of Our Common Natives* (Austin: University of Texas Press, 2009), 11–13; V. Harvard, "Food Plants of the North American Indians," *Bulletin of the Torrey Botanical Club* 22, no. 22 (March 27, 1895): 119.

5. Wild Edibles Database, Honey Locust Tree; Delena Tull, *Edible and Useful Plants of Texas and the Southwest: A Practical Guide* (Austin: University of Texas Press, 1987), 86; and Connie Barlow, *The Ghost of Evolution: Nonsensical Fruit, Missing Partners, and Other Ecological Anachronisms* (New York: Basic Books, 2000), 42–45.
6. Turner, *Remarkable Plants*, 284–87; and Tull, *Edible Plants*, 145–46.
7. Devin S. Stucki, Thomas J. Rodhouse, and Ron J. Reuter, "Effects of Traditional Harvest and Burning on Common Camas (*Camassia quamash*) Abundance in Northern Idaho: The Potential for Traditional Resource Management in a Protected Area Wetland," *Ecology and Evolution* 11, no. 23 (September 1, 2021): 16473–85; Karl W. Kibler and Gemma Mehalchick, "Hunter-Gatherer Resource Acquisition and Use in the Lower Bosque River Basin during the Late Archaic," *Bulletin of the Texas Archaeological Society* 81 (2010): 12; and conversations with O. A. Grant, Tarleton history professor and avocational horticulturalist, 1964.
8. Kibler and Mehalchick, "Hunter-Gatherer Resource," 25, 418.
9. John W. Clark Jr., "Implications of Land and Fresh-Water Gastropods in Archeological Sites," *Journal of the Arkansas Academy of Science* 23, Article 9 (1969):43; I have volunteered with the gaultschool.org since 2016 and have worked the screen as we excavated middens. I have never seen a scorched Rabdotus shell.
10. Wilson W. Crook III, "The Carrollton Phase Archaic: A Redefinition of the Chronology, Composition, and Aerial Distribution of the Early Archaic Horizon along the Trinity River, Texas," *Houston Archaeological Society*, Report Number 35 (2020): 129.
11. Andrew Malof, "Feast or Famine: The Dietary Role of Rabdotus Species Snails in Central Texas" (thesis, University of Texas at San Antonio, 2001), 31.
12. David L. Carlson, ed., "Archeological Investigations along Owl Creek: Results of the Summer Archeological Field School," United States Army Fort Hood, Archeological Resource Management Series, Research Report Number 29, 1997, 16; Malof, "Feast or Famine," 27.
13. Ken Brown, "Archeomalacology: What We Can Learn from Snails," The TARL Blog: The Texas Archeological Research Laboratory, *Texas Beyond History* (March 17, 2015).
14. Kibler and Mehalchick, "Hunter-Gatherer Resource," 317, 337; and Traci Popejoy, Charles R. Randkiev, and Steve Wolverton, "Conservation Implications of Late Holocene Freshwater Mussels Remains of the Leon River in Central Texas," *Hydrobiologia* (October 29, 2016): 1–8.
15. Mussel Collecting, *Texas Beyond History*.
16. Popejoy, Randkiev, and Wolverton, "Conservation Implications," 6.

17. Kibler and Mehalchick, "Hunter-Gatherer Resource," 360–67.
18. Malof, *Feast or Famine*, 36; and my observations along the Bosque River.
19. Robert K. Booth, Stephen T. Jackson, et al., "A Severe Centennial-Scale Drought in Mid-Continental North America 4200 Years Ago and Apparent Global Linkages," *The Holocene* 15, no. 3 (2005): 321–28; Edward W. Herrmann and G. William Monaghan, "Post-Glacial Drainage Basin Evolution in the Midcontinent, North America: Implications for Prehistoric Human Settlement Patterns," *Quaternary International* 511 (2019): 74–76; Uma Shankar, "Cave 'Krem Mawmluh' of Meghalaya Plateau—The Base of the 'Meghalayan Age' and '4.2 ka BP Event' in Holocene (Anthropocene)," *International Journal of Ecology and Environmental Sciences* 47, no. 1 (2021): 49–59; and Michael R. Waters, "Late Quaternary Floodplain History of the Brazos River in East-Central Texas," *Quaternary Research* 43, no. 3 (May 1995): 311–19.
20. Shankar, "Cave 'Krem Mawmluh,'" 49.
21. Kibler and Mehalchick, "Hunter-Gatherer Resource," 304–6; Waco Lake: *Texas Beyond History*; and my observations of Bosque River collections.
22. Karl W. Kibler and Tim Gibbs, "Archeological Survey of 61 Acres along the Bosque River, Waco, McLennan County, Texas," Prewitt and Associates, Inc., Technical Reports, Number 69 (July 2004), 4; Leland C. Bement, *Hunter-Gatherer Mortuary Practices during the Central Texas Archaic* (Austin: University of Texas Press, 1994), 128.
23. Thornton R. Raskevitz, "Phytolith Analysis as a Paleoecological Proxy When Examining Bison Anatomical and Behavioral Changes in the Great Plains" (master's thesis, Oklahoma State University, 2016), 3; and Jon C. Lohse, Stephen L. Black, et al., "Toward an Improved Archaic Radiocarbon Chronology for Central Texas," *Bulletin of the Texas Archeological Society* 85 (2014): 273–78.
24. David Kaniewski, et al., "300-Year Drought Frames Late Bronze Age to Early Iron Age Transition in the Near East: New Paleoecological Data from Cyprus and Syria," *Regional Environmental Change* (December 20, 2018).
25. Lohse and Black, "Toward an Improved Archaic Radiocarbon Chronology," 276–78; Jon C. Lohse, David B. Madsen, et al., "Isotope Paleoecology of Episodic Mid-to-Late Holocene Bison Population Expansions in the Southern Plains, U.S.A.," *Quaternary Science Reviews* 102 (2014): 14–26.
26. The Homeric Grand Solar Minimum (2,800–2,550 years ago) is named for the destruction of the Mycenaean Greek world caused by this event. In North America, the climate was characterized by cooler temperatures, more rain, and an intensively stormy period in the Gulf Coast. Javier, "Impact of the 2,400 Year Solar Cycle on Climate and Human Societies," *Climate,*

Etc., posted on September 20, 2016, https://judithcurry.com/2016/09/20//impact-of-the-2,400-yr-solar-cycle-on-climate-and-human-societies.
27. Lohse, "Toward an Improved Archaic Chronology," 278; and projectile points.net.
28. Waters, "Late Quaternary Floodplain History," 311.
29. Kibler and Mehalchick, "Hunter-Gatherer Resource," 47.
30. Kimberly K. Kvernes, Marie E. Blake, et al., "Re-Location and Updated Recordation of 44 Archeological Sites at Waco Lake, McLennan County, Texas," Prewitt and Associates, Inc., Austin, Texas (June 2000), 14; Stephen M. Carpenter, chapter 11, "Long-Term Subsistence Strategies from the Archaic to Late Prehistoric Times," in "The Siren Site and the Long Transition from Archaic to Late Prehistoric Lifeways on the Eastern Edwards Plateau of Central Texas," *Index of Texas Archeology: Open Access Gray Literature from the Lone Star State* 2013, ed. Stephen M. Carpenter et al., Article 4, 297; and Steve Black and Emily McCuistion, "Life in the Bosque River Basin and Beyond," University of Texas at Austin, 2022.
31. Mary Jo Galindo et al., "Chapter 12, Metric Discrimination of Projectile Points from 41WM1126," in *The Siren Site*, 348–55.

Chapter Six

1. Leland C. Bement, *Hunter-Gatherer Mortality Practices during the Central Texas Archaic* (Austin: University of Texas Press, 1994) 130.
2. Gemma Mehalchick and Karl W. Kibler, "Hunters and Gatherers of the North Bosque River Valley: Excavations at the Baylor, Britton, McMillan, and Higginbotham Sites, Waco Lake, McMillan County, Texas," Prewitt and Associates Inc., Cultural Resources Services, Austin, Texas (July 2008), 154.
3. "Life in the Bosque River Basin and Beyond," *Texas Beyond History*.
4. Stephen Carpenter, Kevin A. Miller, et al., "The Siren Site and the Long Transition from Archaic to Late Prehistoric Lifeways on the Eastern Edwards Plateau of Central Texas," *Index of Texas Archeology: Open Access Gray Literature from the Lone Star State*, 2013, Article 4, 267; Examination of Bosque River collections, and those listed in Kibler and Mehalchick, *Hunters and Gatherers of the North Bosque*, 234.
5. Kibler and Mehalchick, "Hunters and Gatherers of the North Bosque," 252; and "Life along the Bosque River Basin and Beyond," *Texas Beyond History*.
6. Kibler and Mehalchick, *Hunters and Gatherers of the North Bosque*, 13; Ellen Sue Turner and Thomas R. Hester, *A Field Guide to Stone Artifacts of Texas Indians*, 114; Carpenter and Miller, *The Siren Site*, 267.

7. Kibler and Mehalchick, "Hunters and Gatherers of the North Bosque," 368.
8. Kibler and Mehalchick, "Hunters and Gatherers of the North Bosque," 354, 368.
9. Kibler and Mehalchick, "Hunters and Gatherers of the North Bosque," 354, 368; and Jon C. Lohse et al., "Toward an Improved Archaic Radiocarbon Chronology for Central Texas," *Bulletin of the Texas Archeological Society* 85 (2014): 280.
10. Timothy K. Perttula et al., "Cultural Settings of the Leon River Basin," in *Archeological and Geological Test Excavations at Site 41HM61, Hamilton County, Texas*, ed. Robert A. Weinstein, 15–27 (Texas Department of Transportation [TxDOT], 2015), 27; Carpenter and Miller, *The Siren Site*, 344; Grant D. Hall, "Allens Creek: A Study in the Cultural Prehistory of the Lower Brazos River Valley, Texas," *Texas Archeological Survey Research Report*, no. 61, (1981): 112; J. Michael Quigg et al., *Root-Be-Gone (41YN452): Data Recovery of Late Archaic Components in Young County, Texas I* (Austin: Texas Department of Transportation Environmental Affairs Division, TRC Environmental Corporation, 2011).
11. Ulf Buntgen et al., "Cooling and Societal Change during the Late Antique Little Ice Age from 536 to around 660 AD," *Nature Geoscience* 9 (February 8, 2016): 231–36; Ulf Buntgen, Dominique Arseneault, et al., "Prominent Role of Volcanism in Common Era Climate Variability and Human History," *Dendrochronologia* 64 (December 2020).
12. Peter N. Peregrine et al., "Climate and Social Change at the Start of the Late Antique Little Ice Age," *The Holocene* 30, no. 11 (2020): 1643–48; Ulf Buntgen, "Global Wood Anatomical Perspective on the onset of the Late Antique Little Ice Age (LALIA) in the Mid-6th Century CE," *Science Bulletin* 67, no. 22 (November 2022): 2336–44.
13. Robert A. Dull et al., "Radiocarbon and Geologic Evidence Reveal Ilopango Volcano as Source of the Colossal 'Mystery' Eruption of 539–40 CE," *Quaternary Science Reviews* 30 (2019): 1–14; Clive Oppenheimer, *Eruptions That Shook the World* (Cambridge: Cambridge University Press, 2011), 248–51.
14. Matthew Toohey et al., "Climatic and Social Impacts of a Volcanic Double Event at the Dawn of the Middle Ages," *Climatic Chance* 136 (2016): 402.
15. Peter Sarris, "Chapter 18, Climate and Disease," in *A Companion to the Global Early Middle Ages*, ed. Erik Hermans (York, Yorkshire: Arc Humanities Press, 2020), 1–5.
16. Erik Grigg, "'Mole Rain' And Other Natural Phenomena in the Welsh Annals," *Welsh History Review / Cylchgrawn Hanes Cymru* 24, no. 4: 1–40 (2009).
17. Emma Rigby et al., "A Comet Impact in AD 536?" *Astronomy and Geophysics* 45, no. 1 (2004): 1.3–1.6.

18. Mark B. Bush, *Ecology of a Changing Planet* (San Francisco: Benjamin Cummings, 2002), 1–2; J. Siepak et al., "Speciation of Aluminum Released under the Effect of Acid Rain," *Polish Journal of Environmental Sciences* 8, no. 1 (1999): 55–58; G. B. Lawrence et al., "Acid Rain Effects on Aluminum Mobilization Clarified by Inclusion of Strong Organic Acids," *Environmental Science Technology* 41, no. 1 (2007): 1; Theodor Dimitrov, "In the Times of Madness and Fury: Observations on Collective Behavior and Behavioral Deviations in Byzantium in the Age of Justinian Plague (541–750)," *Central and Eastern European Online Library*, no. 27 (2021).
19. Carlos E. Navia, "On the Occurrence of Historical Pandemics during the Grand Solar Minima," *European Journal of Applied Sciences* 2, no. 4 (July 2020): 1–8; Mohammad Hossein Nasirpour et al., "Revealing the Relationship between Solar Activity and COVID-19 and Forecasting of Possible Future Viruses Using Multi-Step Autoregression (MSAR)," *Environmental Science Pollution Research* 28 (July 2021): 38074–84.
20. Monica H. Green et al., "Yersinia Pestis and the Three Plague Epidemics," *The Lancet* 14, no. 10 (October 2014): 918.
21. William Rosen, *Justinian's Flea: Plague, Empire, and the Birth of Europe* (New York: Viking Press, 2007), 105.
22. Rosen, *Justinian's Flea*, 105.
23. R. Barbieri et al., "*Yesinia pestis*: The Natural History of Plague," *Clinical Microbiology Reviews* 34, no. 1 (December 9, 2020); L. Vitello et al., "Preventive Measures against Pandemics from the Beginning of Civilization to Nowadays—How Everything Has Remained the Same over the Millennia," *Journal of Clinical Medicine* 11, no. 7 (April 1, 2022).
24. Sing C. Chew, *The Recurring Dark Ages: Ecological Stress, Climate Changes, and System Transformation* (New York: Altamira Press, 2007).
25. Modern syphilis was brewed up when two ancient lines of treponema bacteria came together with Columbus's debauchery. Fernando Lucas de Melo et al., "Syphilis at the Crossroads of Phylogenetics and Paleopathology," *Neglected Tropical Diseases* 4, no. 1 (2010): 575.
26. Bruce M. Rothschild, Christine Rothschild, and Glen Doran, "Virgin Texas: Treponematosos-Associated Periosteal Reaction 6 Millenia in the Past," *Advances in Anthropology* 1, no. 2 (2011): 15–18; Diane Elizabeth Wilson, "The Paleoepidemiology of Treponematosis in Texas" (PhD diss., University of Texas at Austin, 1998); Maria Ostendorf Smith, "Treponemal Disease in the Middle Archaic to Early Woodland Periods of the Western Tennessee River Valley," *American Journal of Physical Anthropology* 131 (2006): 205–17.
27. Mary J. Adair et al., "Early Maize (Zea mays) in the North American Central Plains: The Microbotanical Evidence," *American Antiquity* 87, no. 2 (2022):

343; Thomas J. Pluckhahn et al., "The History and Future of Migrationist Explanations in the Archeology of the Eastern Woodlands with a Synthetic Model of Woodland Period Migrations on the Gulf Coast," *Journal of Archeological Research* (2020).
28. Navia, *Historical Pandemics*; Sarah Mathena, "Developing a Multistage Model for Treponemal Disease Susceptibility" (master's thesis, Mississippi State University, 2013).
29. Brad Logan, "Late Woodland Feasting and Social Networks in the Lower Missouri River Region," *North American Archeologist* (2022): 1–46.
30. Heinz Wanner et al., "Structure and Origins of Holocene Cold Events," *Quaternary Science Reviews* 30 (July 2011): 1–15; Buntgen, *Societal Change*, 231–36.
31. Logan, *Late Woodland Feasting*, 1; Dale L. McElrath, ed., "Social Evolution or Social Response? A Fresh Look at the 'Good Gray Culture' after Four Decades of Midwest Research," in *Late Woodland Societies: Tradition and Transformation across the Mid-Continent*, ed. Thomas E. Emerson (Lincoln: University of Nebraska Press, 2008), 14–16.
32. Kimberly K. Kvernes et al., "Re-Location and Updated Recordation of 44 Archeological Sites at Waco Lake, McLennan County, Texas," Prewitt and Associates (Austin), US Army Corps of Engineers, Report 127, 16; John H. Blitz, "Adoption of the Bow in Prehistoric North America," *North American Archaeologist* 9, no. 2 (1988): 132.
33. Blitz, *Adoption of the Bow*, 132.
34. Bement, *Hunter-Gatherer Mortuary*, 130; Douglas K. Boyd et al., *Data Recovery Investigations at the Tank Destroyer Site (41CV1378) at Fort Hood, Coryell County, Texas*, Prewitt (Austin), Texas Department of Transportation Environmental Affairs Division, 149 (2014), 133.
35. History is hypothesis, extending many lines of information a little beyond what climatologists, epidemiologists, and archaeologists would be comfortable publishing. The bow empowers the individual. Brigid Sky Grund, "Behavioral Ecology, Technology, and the Organization of Labor: How a Shift from Spear Thrower to Self Bow Exacerbates Social Disparities," *American Anthropologist* 119, no. 1 (2017): 107.
36. Grund, "*Behavioral Ecology*," 106.
37. Quigg, *Root-Be-Gone*, 277; dart and arrowpoints have been found in the same context for a couple of centuries. These campsites are said to be too vague to say anything definite about their association. I have wondered if the two technologies represent separate identities within the same culture.
38. Wilson W. Crook III and Mark D. Hughston, "The Late Prehistoric of the East Fork of the Trinity River," *CRHR Research Reports* 2, Article 1 (2016): 6;

R. A. Ricklis, "Archaeology and Bioarchaeology of the Buckeye Site (41VT98), Victoria County, Texas," Coastal Environments, Inc. and U.S. Army Corps of Engineers, Galveston District (2012), 28; Quigg, *Root-Be-Gone*, 16. I have found Darl points among arrowpoints that were less than an inch in length. Hypotheses based on field observations and discussions with geoarchaeologist and veteran chert knapper, Charles Frederick.

39. I have examined the upper half of the Bosque River's gravel bars and found them to be chert-poor. The lower part of the river contains larger cobbles and spalls of Edwards fine cherts, but these sources would have been closed to low-status peoples.

40. Grund, *Behavioral Ecology*, 119; Carpenter and Miller, *The Siren Site*, 353–55; Alston V. Thoms, ed., "Archeological Survey at Fort Hood, Texas Fiscal Years 1991 and 1992: The Cantonment and Belton Lake Periphery Areas," United States Army Fort Hood Archeological Resource Management Series, Research Report No. 27 (1993), 44; my field observations.

41. Edward B. Jelks, *The Kyle Site: A Stratified Central Texas Aspect Site in Hill County, Texas, Archeology Series, No. 5* (Austin: The University of Texas Press, 1962), 69; Danny L. Hamilton, *Prehistory of the Rustler Hills Granado Cave* (Austin: University of Texas Press, 2001), 166; observations of private Trans-Pecos collections; the hardwood portion of arrows along the Bosque River must have been rough leaf dogwood (*Cornus drummondii*).

42. The behavior of arrows made from the right or left wing is of European origin, and this practice is assumed to have been a concern along the Bosque River by me; Saxton T. Pope, *Bows and Arrows* (Berkeley: University of California Press, 1962), 49–50.

43. Leslie L. Bush. "Evidence for a Long-Distance Trade in Bois d'arc Bows in 16th Century Texas (*Maclura pomifera* Moraceae)," *Journal of Texas Archeology and History* 1 (2014): 51–69; the bois d'arc tree was unavailable along the Bosque River until historic times. After a flood in the 1970s, Stephenville collector Avis Delk saw the tip of a bow in a context of packed leaves under a rock ledge near Stephenville City Park. He said that he excavated the entirety of the leaf material hoping to find an arrow. I examined the bow and noted that it was polished and had grooves along its four-foot length. Avis tried to string the bow and broke it. I think it was lost when his house burned.

44. Stephen A. Hall et al., "New Correlation of Stable Carbon Isotopes with Changing Late-Holocene Fluvial Environments in the Trinity River Basin of Texas, USA," *The Holocene* 22 (2012): 5; Stephen A. Hall. "Channel Trenching and Climatic Change in the Southern U.S. Great Plains," *Geology* 18 (1990): 342–45; Rebecca Taormina, Lee Nordt, et al., "Late Quaternary Alluvial History of the Brazos River in Central Texas," *Quaternary International*

631 (September 10, 2022): 34–46; Holly A. Meier, Lee C. Nordt, et al., "Late Quaternary Alluvial History of the Middle Owl Creek Drainage Basin in Central Texas: A Record of Geomorphic Response to Environmental Change," *Quaternary International* 30 (2013): 1–18; Carpenter and Miller, *The Siren Site*, 303; and the marshy nature of the Bosque River and camas meadows is based on discussions with Charles Frederick, and my inferences.

45. Quigg, "Root-Be-Gone," 304; Todd L. VanPool, "The Survival of Archaic Technology in an Agricultural World: How the Atlatl and Dart Endured in the North American Southwest," *Journal of Southwestern Anthropology and History* 71, no. 4 (2006): 429–52.

46. Bradley J. Vierra, Nicholas Chapin, et al., "Another Look at Expedient Technologies, Sedentism, and the Bow and Arrow," *Journal of Southwestern Anthropology and History* 86, no. 4 (2020): 7; Kerry Nichols, "Late Woodland Cultural Adaptions in the Lower Missouri River Valley: Archery, Warfare, and the Rise of Complexity" (PhD diss., University of Missouri–Columbia, 2015), 53; Julianne E. Tarabek. "What's the Point: The Transition from Dart to Bow in the Eastern Plains" (master's thesis, University of Kansas, 2013), 40.

47. Carpenter and Miller, "The Siren Site," 349, 358–59.

48. Edward B. Jelks, *The Kyle Site: A Stratified Central Texas Aspect Site in Hill County, Texas, Archeology Series, no. 10* (Austin: The University of Texas, 1962), 28.

49. Cory Leahy, "Medieval Warm Period Not So Random," *UT News*, November 11 2010.

50. William C. Foster, *Climate and Culture Change in North America, AD 900–1600* (Austin: University of Texas Press, 2012), 16, 156.

51. Kimberly K. Kvernes, Marie E. Blake, et al., "Relocation and Updated Recordation of 44 Archeological Sites at Waco Lake, McLennan County, Texas," *US Army 20000628* (2000): 16; Chuck Hixson, "Graham-Applegate Rancheria: What Is the Austin Phase," *Texas Beyond History*, 2001; Timothy K. Perttula, Duncan McKinnon, Scott Hammerstedt, "The Archaeology, Bioarchaeology, Ethnography, Ethnohistory, and History Bibliography of the Caddo Indian Peoples of Arkansas, Louisiana, Oklahoma, and Texas," *Index of Texas Archeology: Open Access Gray Literature for Lone Star State*, 2021, Article 1 (2021). While I was a student at Tarleton College in the early 1960s I was given two especially well-crafted Scallorn points that were said to have been in a skeleton bulldozed up near a rock shelter close to Cranfills Gap, Texas.

52. Charles Hixson and Buddy Whitley, "Early and Late Toyah-Period Occupations within Area C of the Baker Site on the Northeastern Edwards Plateau, Central Texas," *Journal of Texas Archeology and History* 7 (2023): 63–64; Timothy K. Perttula, "Prairie Caddo Sites in Coryell and McLennan Counties

in Central Texas," *Index of Texas Archeology: Open Access Gray Literature from the Lone Star State* Article 102 (2016): 42; Bradie Dean, "Caddo Artifacts in Central Texas: A Proposed Trade Connection" (master's thesis, Baylor University, 2020), 1, 53, 62.

53. Dean, "Caddo Artifacts," 52–62; and Ross C. Fields. "The Prairie Caddo Model and the J. B. White Site," *Index of Texas Archeology: Open Access Gray Literature from the Lone Star State 2017*, Article 43 (2017): 17.

Chapter Seven

1. This description is drawn from studies in organic carbon from the West Range alluvium, pollen, molluscan, faunal, and sedimentary structures. Lee C. Nordt, "Archeological Geology of the Fort Hood Military Reservation Fort Hood, Texas," *United States Army, Fort Hood, Archeological Resource Management Series, Report 25* (1992): 67; Stephen A. Hall et al., "New Correlation of Staple Carbon Isotopes with Changing Late-Holocene Fluvial Environments in the Trinity River Basin of Texas, USA," *The Holocene* 22, no. 5 (2011): 546; Christopher Linz et al., "Archeological Testing at 41TR170, along the Clear Fork of the Trinity River, Tarrant County, Texas," *Texas Department of Transportation, Environmental Affairs Division, Number 348* (2008); El Niño suppresses Atlantic hurricanes, and La Niña has the opposite effect. Arthur L. Bishop, "Flood Potential of the Bosque Basin," *Baylor Geological Studies*, no. 33 (1977): 35; Stephen A. Hall, "Channel Trenching and Climate Change in the Southern U.S. Great Plains," *Geology* 18 (1990): 344; Kim M. Cobb et al., "El Nino/Southern Oscillation and Tropical Pacific Climate during the Last Millennium," *Nature* 424 (2003): 271–76; J. Michael Daniels et al., "Alluvial Stratigraphic Evidence for Channel Incision during the Medieval Warm Period on the Central Great Plains," *Holocene* 15, no. 5 (2005): abstract; and examinations of Bosque River sedimentation with Charles Frederick in 2018.

2. Larry V. Benson, Michael S. Berry et al., "Possible Impact of Early-11th, Middle-12th, and Late-13th-Century Droughts on Western Native Americans and the Mississippian Cahokians," *Quaternary Science Reviews* 26 (2007): 336–50; Michael R. Waters, Joshua L. Keene, et al., "Late Quaternary Geology, Archaeology, and Geoarchaeology of Hall's Cave, Texas," *Quaternary Science Reviews* 274 (December 15, 2021): 4.2; the massive erosional event that washed away the West Range alluvium is thought to have been caused by rainstorms falling on recent landscape made barren by drought. There is new research suggesting that low solar winds, a product of solar minima (like the Wolf

Minimum) can cause super-thunderstorms, which would help explain such a floodplain-changing episode; Rusin P. Prikry, V. Prikryl, et al., "Heavy Rainfall, Floods, and Flash Floods Influenced by High-Speed Solar Wind Coupling to the Magnetosphere-Ionosphere System," *Annales Geophysicae* 39, no. 4 (April 2021): 769–93.

3. Francois Lapointe and Raymond S. Bradley, "Little Ice Age Abruptly Triggered by Intrusion of Atlantic Waters into the Nordic Seas," *Science Advances* 7, no. 51 (December 2021); Edouard Bard and Martin Frank, "Climate Change and Solar Variability: What's New Under the Sun? *Earth and Planetary Science Letters* 248 (2006): 1–14.

4. Ulf Buntgen, Dominique Arseneault, et al., "Prominent Role of Volcanism in Common Era Climate Variability and Human History," *Dendrochronogia* 64 (December 2020); William C. Foster, *Climate and Culture Change in North America, AD 900–1600* (Austin: University of Texas Press, 2012), 64.

5. Erickson and Boyd, *Porch Talk*, 95; Christopher Lintz, "Texas Panhandle-Pueblo Interactions from the Thirteenth Century through the Sixteenth Century," in *Farmers, Hunters, and Colonists: Interaction between the Southwest and the Southern Plains*, ed. Kathrine A. Spielman (Tucson: University of Arizona Press, 1991), 177; Martin P. R. Magne and R. G. Matson, "Moving On: Expanding Perspectives on Athapaskan Migration," *Canadian Journal of Archaeology* 34 (2010): 229.

6. Laura Kozuch, "Olivella Beads from Spiro and the Plains," *American Antiquity* 67, no. 4 (2002): 698; Steve Black, "Spiro and the Arkansas Basin, Caddo Fundamentals," *Texas Beyond History* (August 2003).

7. Thomas A. Jennings and Ashley M. Smallwood, "Clovis and Toyah: Convergent Blade Technologies on the Southern Plains Periphery of North America," in *Convergent Evolution in Stone-Tool Technology*, ed. Briggs Buchanan, Michael J. O-Brien, et al. (Cambridge, MA: MIT Press, 2018), 229–53; Jon Budd and John E. Dockall, et al., *Testing and Data Recovery Excavations at the Jayroe Site, (41HM51), Hamilton County, Texas*, Prewitt and Associates, Inc. (Austin), Cultural Resources of Texas, Report 187, I-100, I-123 (2020).

8. Wilson W. Crook and Mark D. Hughston, "The Late Prehistoric of the East Fork of the Trinity River," *CRHR Research Reports* 2, Article 1 (2016); LeRoy Johnson, *The Life and Times of Toyah-Culture Folk* (Austin: Texas Department of Transportation and Texas Historical Commission, Office of the State Archeologist Report, 38 (1994), 89; Hixson and Whitley, "Early and Late Toyah within Area C of the Baker Site on the Northeastern Edwards Plateau, Central Texas," *Journal of Texas Archeology and History* 7, Article 2 (2023): 60–61; William C. Foster, *Climate and Culture Change in North America, AD 900–1600* (Austin: University of Texas Press, 2012), 114.

9. Foster, *Climate and Culture Change*, 113.
10. Dockall, Fields, et al., *Testing and Data Recovery Excavations at the Jayroe Site (41HM51), Hamilton County, Texas (Waco District, CSJ No. 0909-29-030)* (Austin, Texas Department of Transportation, 2020), 24–25; Hixson and Whitley, "Early and Late Toyah," 57–58.
11. W. W. Newcomb Jr., *The Indians of Texas: From Prehistoric to Modern Times* (Austin: University of Texas Press, 1961), 23; Mildred Mott Wedel, "The Wichita Indians in the Arkansas River Basin," *Plains Indians Studies: A Collection of Essays in Honor of John C. Ewers and Waldo R. Wedel* 30 (1982): 118.
12. F. Todd Smith, *The Wichita Indians: Traders of Texas and the Southern Plains, 1540–1845* (College Station: Texas A&M University Press, 2000), 3, 6.
13. Natalie G. Mueller, "The Occurrence of a Newly Described Domesticate in Eastern North America: Adena/Hopewell Communities and Agricultural Innovation," *Journal of Anthropological Archaeology* 49 (March 2018): abstract; Natalie G. Mueller, "Seeds as Artifacts of Communities of Practice: The Domestication of Knotweed in Eastern North America" (PhD diss., Washington University in St. Louis, 2017), 6–7.
14. John P. Hart and William A. Lovis, "Reevaluating What We Know about the Histories of Maize in Northeastern North America: A Review of Current Evidence," *Journal of Archaeological Research* 21 (2013): 195.
15. Natalie G. Mueller, Gayle J. Fritz, et al., "Growing the Lost Crops of Eastern North America's Original Agricultural System," *Nature Plants* 3 (2017): 1; Kindscher, *Edible Wild Plants of the Prairie*, 139–41.
16. Mueller, *Lost Crops*, 1; Mueller, *Seeds as Artifacts*, 7; and Kindscher, *Edible Wild Plants*, 233.
17. John P. Hart, "Evolving the Three Sisters: The Changing Histories of Maize, Bean, and Squash in New York and the Greater Northeast," in *Current Northeast Paleoethnobotany II*, ed. John P. Hart et al. (New York: State Museum Bulletin 512, 2008), 1–4; John P. Hart, "Maize Agriculture Evolution in the Eastern Woodlands of North America: A Darwinian Perspective," *Journal of Archaeological Method and Theory* 6, no. 2 (1999): 140.
18. Guy Gibbon, "Lifeways through Time in the Upper Mississippi River Valley and Northeastern Plains," 332, in *The Oxford Handbook of North American Archaeology*, ed. Timothy Pauketat (Oxford: Oxford University Press, 2012).
19. G. William Monaghan, Timothy M. Shilling, et al., "The Age and Distribution of Domesticated Beans (*Phaseolus vulgaris*) in Eastern North America: Implications for Agricultural Practices and Group Interactions," *Midwest Archaeological Conference, Occasional Papers*, no. 1 (2014): 33–34; Deni J. Seymour, "Comments on Genetic Data Relating to Athapaskan Migrations: Implications of the Malhi Study for the Southwestern Apache and Navajo,"

American Journal of Physical Anthropology 139 (2009): 281–83; Erickson and Boyd, *Porch Talk*, 103–5.
20. Severin Fowles, "The Pueblo Village in the Age of Reformation (AD 1300–1600)," 632–33, in *The Oxford Handbook of North American Archaeology*, ed. Timothy R. Pauketat (Oxford: Oxford University Press, 2012).
21. Juliana Barr, "There Is No Such Thing as 'Prehistory': What the Longue Duree of Caddo and Pueblo History Tells Us about Colonial America," *William and Mary Quarterly* 74, no. 2 (April 2017): 207.
22. Barr, *There's No Such Thing as Prehistory*, 208, 228–29. When Europeans encountered the Wichita they found that they were still relatively democratic.
23. Kaitlyn Elizabeth Davis, "The Ambassador's Herb: Tobacco Pipes as Evidence for Plains-Pueblo Interaction, Interethnic Negotiation, and Ceremonial Exchange in the Northern Rio Grande" (master's thesis, University of Colorado at Boulder, 2017), 13–15, 101–5; Christopher B. Roding. "Cherokee Towns and Calumet Ceremonialism in Eastern North America," *American Antiquity* 79, no. 4 (2014): 425–43.
24. Bennett Harrison Dampier, "The Moth and the Moonflower: Datura and Hawk Moth Iconography across Ancient America" (master's thesis, Texas State University, 2022), 142–43.
25. Anna Hardwick Gayton, "The Narcotic Plant Datura in Aboriginal American Culture" (PhD diss., University of California at Berkley, 1926), 42–44.
26. Frederick H. Douglas, "The Wichita Indians and Allied Tribes: Waco, Towakoni, and Kichi," *Denver Art Museum*, no. 40 (January 1932): 2.
27. George A. Dorsey, *The Mythology of the Wichita* (Carnegie Institution of Washington, DC, 1904), 25.
28. Rudolph C. Troike, "The Origins of Plains Mescalism," *American Anthropologists* (1962): 956, 958; Gayton, *The Narcotic Plant Datura*, 42–50.
29. The Jumano Indians lived at La Junta, the junction of the northward-flowing Concho River and the Rio Grande across from Presidio. They were winter traders and summer farmers who connected the southern Pueblos with the Caddoan villages. Oral history has it that the Jumanos held trade fairs at Comanche Peak, near Granbury, with Apaches and Caddo/Wichitas meeting peacefully to trade. Dan M. Young, "Identification of the Jumano Indians" (master's thesis, Sul Ross University, 1970), 18, 28, 43; Troike, "The Origins of Plains Mescalism," 946–63; Turner, *A Field Guide to Artifacts*, 156.
30. Cecile Elkins Carter, *Caddo Indians: Where We Come From* (Norman: University of Oklahoma Press, 1995), 135.
31. F. Todd Smith, *The Caddo Indians: Tribes at the Convergence of Empires, 1542–1854* (College Station: Texas A&M University Press, 1995), 12; Charles M. Hudson, ed., *Black Drink: A Native American Tea* (Athens: University of

Georgia Press, 1979), 2–5; John C. Ewers, ed., *The Indians of Texas in 1830 by Jean Louis Berlandier* (Washington, DC: Smithsonian Institution Press, 1969), 94–95.
32. Hudson, *Black Drink*, 3.
33. Ewers, *The Indians of Texas*, 95.

Chapter Eight

1. F. Todd Smith, "Wichita Locations and Population, 1719–1901," *Plains Anthropologist* 53, no. 208 (2008): 407–14; Rodolfo Acuna-Soto, David W. Stahie, et al., "Megadrought and Megadeath in 16th Century Mexico," *Emerging Infectious Diseases* 4 (2002): 360–62; and Ashild J. Vagene, Michael G. Campana, et al., "*Salmonella enterica* Genomes Recovered from Victims of a Major 16th Century Epidemic in Mexico," *Nature, Ecology, and Evolution* 2, no. 2 (2017): 520–28.
2. Gary Clayton Anderson, *The Indian Southwest, 1580–1830: Ethnogenesis and Reinvention* (Norman: University of Oklahoma Press, 1999), 57, 108, 165.
3. R. G. Robertson, "Rotting Face: Smallpox and the American Indian," *Journal of American History* 90, no. 1 (2003): 221–22; Stephen M. Perkins and Timothy Baugh, "Protohistory and the Wichita," *Plains Anthropologist* 53, no. 208 (November 2008): 383; and Robert C. Vogel, "Paul Laffite: A Borderland Life," *East Texas Historical Journal* 41, no. 1 (2003): 19.
4. Solar Minimums facilitate the development of new forms of diseases. Carlos E. Navia, "On the Occurrence of Historical Pandemics during the Grand Solar Minima," *European Journal of Applied Physics* 2, no. 4 (July 2020): 1–8; and Anderson, *The Indian Southwest*, 57; Stephen M. Perkins and Timothy Baugh, "Protohistory and the Wichita," *Plains Anthropologist* (November 2008): 381–89; and the Wichitas preferred the lighter French flintlocks over the heavy Spanish military muskets; Timothy G. Baugh and Jay C. Blaine, "Enduring the Violence: Four Centuries of Kirikir'i·s Warfare," *Plains Anthropologist* 62, no. 242 (2017): 99–132.
5. W. W. Newcomb Jr., *The Indians of Texas: From Prehistoric to Modern Times* (Austin: University of Texas Press, 1961), 249.
6. Stephen M. Perkins, Susan C. Vehik, and Richard R. Drass, "The Hide Trade and Wichita Social Organization: An Assessment of Ethnological Hypotheses Concerning Polygyny," *Plains Anthropologist* 53, no. 208 (2008): 432–43.
7. Weston La Barre, *The Peyote Cult* (Norman: University of Oklahoma Press, 2012), 14, 60–61; Rudolph C. Troike, "The Origins of Plains Mescalism," *American Anthropologist* 64, no. 59 (1962): 58; and Michael Pollan, *The Botany*

of Desire: A Plants-Eye View of the World (New York: Random House, 2001), 142–43.
8. F. Todd Smith, *The Caddo Indians: Tribes at the Convergence of Empires, 1542–1854* (College Station: Texas A&M University Press, 1995), 8; Timothy G. Baugh and Jay C. Blaine, "Enduring the Violence: Four Centuries of Kirikir'i·s Warfare," *Plains Anthropologist* 62, no. 242 (2017), 111–12; and Elizabeth A. H. John, *Storms Brewed in Other Men's Worlds: The Confrontation of Indians, Spanish, and French in the Southwest, 1540–1795* (Norman: University of Oklahoma Press, 1996), 305.
9. Perkins, Vehik, and Drass, "The Hide Trade," 436; Mark Edward Firmin, "For the Pleasure of the People: A Centennial History of William Cameron Park, Waco, Texas" (PhD thesis, Baylor University, 2009), 2–3; and Karl W. Kibler and Tim Gibbs, "Archeological Survey of 61 Acres along the Bosque River, Waco, McLennan County, Texas," Prewitt and Associates (Austin), Technical Paper, no. 69 (2004): 5.
10. Frank H. Watt. "The Waco Indian Village and Its Peoples," *Texana* 6, no. 3 (Fall 1968): 203; and Newcomb, *Indians of Texas*, 254; J. D. Legare, ed., *Southern Agriculturalist*, vol. 3, 1830, Charleston: A. E. Miller, 1830, 240; B. T. Galloway (Chief of Bureau), "Agricultural Varieties of the Cowpea and Immediately Related Species," US Department of Agriculture, no. 229; today this variety is called Bisbee, which I have grown on trellises; I was disappointed at how few seeds were produced in relation to the number of thick vines; Gary Paul Nabhan, *Native Seeds/Search*, nativeseeds.org.
11. J. H. Connell (director), *The Peach* (College Station: Texas Agricultural Experiment Station, no. 39, July 1896): 805; and Peter Hatch. "We Abound in the Luxury of the Peach," *Twinleaf* 10 (1998): 9–11.
12. Watt, *Waco Indian Village*, 203.
13. Baugh and Blaine, "Enduring Violence," *113*.
14. Anderson, *The Indian Southwest*, 123.
15. Robert S. Weddle, *The San Saba Mission: Spanish Pivot in Texas* (Austin: University of Texas Press, 1964), 68–71, 120–27.
16. Anderson. *The Indian Southwest*, 99.
17. Anderson, *The Indian Southwest*, 159.
18. Smith, *The Wichita Indians*, 44.
19. F. Todd Smith, *From Dominance to Disappearance: The Indians of Texas and the Near Southwest, 1786–1859* (Lincoln: University of Nebraska Press, 2005), 31.
20. Kathrine Turner-Pearson, "The Stone Site: A Waco Indian Village Frozen in Time," *Plains Anthropologist* 53, no. 208 (2008): 565–75.
21. Smith, *The Wichita Indians*, 46.

22. Elizabeth A. Fenn, *Pox Americana: The Great Smallpox Epidemic of 1775–82* (New York: Hill and Wang, 2001), 16, 21, 27; and Smith, *The Caddo Indians*, 74.
23. Fenn, *Pox Americana*, A14, 42–43, 46–55; and Clark Spencer Larsen, *Skeletons in Our Closet: Revealing Our Past through Bioarchaeology* (Princeton, NJ: Princeton University Press, 2000), 136–37.
24. Smith, *The Wichita Indians*, 65.
25. Smith, *The Caddo Indians*, 75.
26. Carter, *Caddo Indians*, 219–22.
27. J. Frank Dobie, *The Longhorns* (Austin: University of Texas Press, 1982), 10; Chipman, *Spanish Texas*, 213–14; Smith, *The Wichita Indians*, 90–91; and Anderson, *The Indian Southwest*, 189–90.
28. David La Vere, *The Texas Indians* (College Station: Texas A&M University Press, 2004), 163.
29. Qian Peng, Ian R. Gizer, et al., "Associations between Genomic Variants in Alcohol Dehydrogenase (ADH) Genes and Alcohol Symptomatology in American Indians and European Americans: Distinctions and Convergence," *Alcoholism: Clinical and Experimental Research* (October 1, 2018): 1695–704.
30. Mary-Ann Enoch and Bernard J. Albaugh, "Genetic and Environmental Risk Factors for Alcohol Use Disorders in American Indians and Alaskan Natives," *America Journal of Addictions* (August 2017): 461–68.
31. La Vere, *Texas Indians*, 163.
32. Randolph B. Campbell, *Gone to Texas: A History of the Lone Star State* (New York: Oxford University Press, 2003), 100–105.
33. Smith, *Dominance to Disappearance*, 131.
34. Campbell, *Gone to Texas*, 115; and Anderson, *Conquest of Texas*, 54–55.
35. Thomas F. Seiter, "Karankawas: Reexamining Texas Gulf Coast Cannibalism" (thesis, University of Houston, 2019), 59–61, 157.
36. Smith, *From Dominance to Disappearance*, 131.
37. Smith, *Conquest of Texas*, 54.
38. David Hackett Fischer, *Albions's Seed: Four British Folkways in America* (New York: Oxford University Press, 1989), 207–55.
39. Fischer, *Albion's Seed*, 605–6.
40. Fischer, *Albion's Seed*, 650–51, 671, 680; and James Webb, *Born Fighting: How the Scots-Irish Shaped America* (New York: Broadway Books, 2004).
41. T. R. Fehrenbach, *Lone Star: A History of Texas and the Texans* (New York: Wing Books, 1968), 95–96; and Fischer, *Albion's Seed*, 655–57.
42. Mary Whatley Clarke, *Chief Bowles and the Texas Cherokees* (Norman: University of Oklahoma Press, 1971), 28; Joel Poinsett is also remembered for introducing the red-leafed plant to the United States; Jess Righthand, "How

Joel Poinsett, the Namesake for the Poinsettia, Played a Role in Creating the Smithsonian," December 6, 2010, smithsonianmag.com.
43. Anderson, *The Conquest of Texas*, 74.
44. J. W. Wilbarger, *Indian Depredations in Texas: Reliable Accounts* (Austin: Hutchings Printing House, 1889), 177–78; Campbell, *Gone to Texas*, 116; Frank H. Watt, "The Waco Indian Village and Its Peoples," *Texana* 6, no. 3 (Fall 1968): 218; Anderson, *Conquest of Texas*, 74–75; and Smith, *The Wichita Indians*, 123.
45. Wilbarger, *Indian Depredations*, 178–79; Watt, "The Waco Indian Village and Its Peoples," 218–19; and Anderson, *Conquest of Texas*, 75.
46. Anderson, *Conquest of Texas*, 75.
47. Stephen L. Moore, *Savage Frontier: Rangers, Riflemen, and Indian Wars in Texas: Volume IV, 1835–1837* (Denton: University of North Texas Press, 2010), 29.
48. Barbara Alice Mann, *The Tainted Gift: The Disease Method of Frontier Expansion* (Santa Barbara, CA: Praeger, 2009); and Robertson, *Rotting Face*, 40–41, 283.
49. Jane Reinhiller, "Holding on to Culture: The Effects of the 1837 Smallpox Epidemic on Mandan and Hidatsa," *Butler Journal of Undergraduate Research* 4 (2018): 206–7; and Robertson, *Rotting Face*, 284–87.
50. Newcomb, *Indians of Texas*, 276.

Chapter Nine

1. H. G. Perry, *Grand Ol' Erath: The Saga of a Texas West Cross Timbers County* (Stephenville, TX: Stephenville Printing, 1974), 7–8; Pekka Hämäläinen, *The Comanche Empire* (New Haven, CT: Yale University Press, 2008), 1.
2. Hämäläinen, *Comanche Empire*, 303, 340; Connie A. Woodhouse and Jonathan T. Overpeck, "2000 Years of Drought Variability in the Central United States," *Bulletin of the American Meteorological Society* 79, no. 12 (December 1998): 2696.
3. Hämäläinen, *Comanche Empire*, 21; Robert K. Booth, John E. Kutzbach, et al., "A Reanalysis of the Relationship between Strong Westerlies and Precipitation in the Great Plains and Midwest Regions of North America," *Climate Change* 76 (2006): 436.
4. Robert L. Beschta and William J. Ripple, "Yellowstone's Prehistoric Bison: A Comment on Keigley (1019)," *Rangelands* 41, no. 3 (2019): 149–51; William C. Foster, *Climate and Culture Change in North America, AD 900–1600* (Austin: University of Texas Press, 2012), 163; Hämäläinen, *Comanche Empire*, 22.

5. G. W. Tomanek and G. K. Hullet, "Effects of Historical Droughts on Grassland Vegetation in the Central Great Plains," in *Pleistocene and Recent Environments of the Central Great Plains*, ed. Wakefield Dort Jr. and J. Knox Jones Jr. (Lawrence: University Press of Kansas, 1970): 203–10; Cody Newton, "Towards a Context for Late Precontact Culture Change: Comanche Movement Prior to Eighteenth Century Spanish Documentation," *Plains Anthropologist* 56, no. 217 (2011): 5; J. E. Weaver, "A Seventeen-Year Study of Plant Succession in Prairie," *American Journal of Botany* 41, no. 1 (January 1954): 31–38.
6. Newton, "Towards a Context for Late Precontact Culture," 5–6; William E. Moore, "A Guide to Texas Arrowpoints," Brazos Valley Research Associates, *Contributions in Archeology*, no. 6 (2015): 51.
7. Elizabeth A. John, *Storms Brewed in Other Men's Worlds: The Confrontation of Indians, Spanish, and French in the Southwest, 1540–1795* (Norman: University of Oklahoma Press, 1996), 101, 112.
8. Susan M. Deeds, "Colonial Chihuahua: Peoples and Frontiers in Flux," in *New Views of Borderland History*, ed. Robert G. Jackson (Albuquerque: University of New Mexico Press, 1998), 31.
9. Newton, "Towards a Context for Late Precontact Culture," 4.
10. Anderson, *The Indian Southwest*, 56–57; and Newton, "Towards a Context for Late Precontact Culture," 10–11; Hämäläinen, *Comanche Empire*, 18; and Newton, "Towards a Context for Late Precontact Culture," 11–13.
11. Smith, *From Dominance to Disappearance*, 8–9.
12. Hämäläinen, *Comanche Empire*, 1–3.
13. Pekka Hämäläinen, *Indigenous Continent: The Epic Contest for North America* (New York: Liveright Publishing, 2022), 412; J. Frank Dobie, *The Longhorns* (Austin: University of Texas Press, 2005), 7–13.
14. When cattle were fenced away from the river in the 1880s, thorny shrubs crowded out these berries. Dan Young, *A Calendar of Erath County History, 1986* (Stephenville, TX: Vanderbilt Street Press, 1986); Tull, *A Practical Guide to Edible and Useful Plants*, 198, 219; Gustav Carlson and Volney Jones, "Some Notes on Uses of Plants by the Comanche Indians," *Papers of the Michigan Academy of Science, Arts, and Letters* 25 (1939): 517–42; personal communication (July 15, 2020) with Leslie Bush.
15. Stanley Noyes, *Los Comanches: The Horse People, 1751–1845* (Albuquerque: University of New Mexico Press, 1993), xxv; Earnest Wallace and E Adamson Hoebel, *The Comanches: Lords of the South Plains* (Norman: University of Oklahoma Press, 1952), 41.
16. Wallace and Hoebel, *The Comanches*, 41–43.

17. Wallace and Hoebel, *The Comanches*, 43; Karen Jones, "The Story of Comanche: Horsepower, Heroism, and the Conquest of the American West," *War and Society* 36, no. 3 (2017): 160.
18. Hämäläinen, *Indigenous Continent*, 413; Hämäläinen, *Comanche Empire*, 2; Jones, "Story of Comanche," 160.
19. Organizations founded on innovation experience a period of success before becoming atrophied; T. R. Fehrenbach, *Comanches: The Destruction of a People* (New York: Da Capo Press, 1994), 89; Anderson, *Indian Southwest*, 233.
20. Hämäläinen, *Comanche Empire*, 240; Eric Tippeconnic, "God Dogs and Education: Comanche Traditional Cultural Innovation and Three Generations of Tippeconnic Men" (PhD diss., University of New Mexico, 2016), 53.
21. Hämäläinen, *Comanche Empire*, 242–43; Dan Flores, ed., *Journal of an Indian Trader: Anthony Glass and the Texas Trading Frontier; 1790–1810* (College Station: Texas A&M University Press, 1985), 81; Hämäläinen, *Comanche Empire*, 245.
22. Anderson, *Indian Southwest*, 230.
23. Newcomb, *Indians of Texas*, 162.
24. Flores, *Journal of an Indian Trader*, 60; Ewers, *The Indians of Texas*, plate 4; Turner, *Remarkable Plants of Texas*, 45; Arden Jean Schuetz, *People-Events and Erath County, Texas* (Stephenville, TX: Ennis Favors, 1970).
25. Newcomb, *Indians of Texas*, 165; Tull, *A Practical Guide to Edible and Useful Plants*, 198; Mary Englar, *The Comanches: Nomads of the Southern Plains* (Mankato, MN: Capstone Press, 2004), 12–13.
26. Newcomb, *Indians of Texas*, 162; David La Vere, *Life among the Texas Indians: The WPA Narratives* (College Station: Texas A&M University Press, 1998), 16.
27. Wallace and Hoebel, *The Comanches*, 61–62.
28. Bernd Heinrich, *Mind of the Raven: Investigations and Adventures with Wolf Birds*, (New York: HarperCollins, 2009), 39; Lawrence Kilham, *The American Crow and the Common Raven* (College Station: Texas A&M University Press, 1989), 183; Daniel Stahler and Brend Heinrich, "Common Ravens, *Corvus corax*, Preferentially Associate with Grey Wolves, *Canis lupus*, as a Foraging Strategy in Winter," *Animal Behavior* 64 (2002): 283–90.
29. Wallace and Hoebel, *The Comanches*, 62.
30. Joaquín Rivaya-Martinez, "A Different Look at Native American Depopulation: Comanche Raiding, Captive Taking, and Population Decline," *Ethnohistory* 61, no. 3 (2014): 392.
31. The Mormon explanation for indigenous enslavement was based on the Book of Mormon's story about a North American Israelite group of degraded Lamanites—American Indian children that could be redeemed through

adoption and a twenty-year indenture. Andres Resendez, *The Other Slavery: The Uncovered Story of Indian Enslavement in America* (Boston: Mariner Books, 2017), 268–73; Michael Kay Bennion, "Captivity, Adoption, Marriage, and Identity: Native American Children in Mormon Homes, 1847–1900," UNLV Theses, Dissertations, Professional Papers, and Capstones, 2012.

32. Hämäläinen, *Comanche Empire*, 255; Pauline Turner Strong, "Transforming Outsiders: Captivity, Adoption, and Slavery Reconsidered," in *A Companion to American Indian History*, ed. Philip J. Deloria and Neal Salisbury (Hoboken, NJ: Wiley-Blackwell, 2004), 351.
33. Hämäläinen, *Comanche Empire*, 254; Rivaya-Martinez, *A Different Look at Native American Depopulation*, 396.
34. Hämäläinen, *Comanche Empire*, 254–55. This was the same smallpox epidemic that devastated the British army in the East; Hämäläinen, *Indigenous Continent*, 415.
35. Tippeconnic, "GodDogs," 61–62.
36. Ewers, *Indians of Texas*; Wallace and Hoebel, *The Comanches*, 142.
37. Wallace and Hoebel, *The Comanches*, 142.
38. Daniel E. Moerman, *Native American Medicinal Plants: An Ethnobotanical Dictionary* (Portland, OR: Timber Press, 2009), 277; Kindsher, *Medicinal Wild Plants of the Prairie*, 143.
39. Wallace and Hoebel, *The Comanches*, 142; Turner, *Remarkable Plants of Texas*, 292; Moerman, *Native American Plants*, 507.
40. Hämäläinen, *Comanche Empire*, 243–44; Noah Smithwick, *The Evolution of a State, or, Recollections of Old Texas Days*, ed. Nanna Smithwick Donaldson (Austin: Gammel Book, 1900), 135; sometimes severely wounded Comanches would kill themselves; Anderson, *The Indian Southwest*, 234.
41. Flores, *Journal*, 133; Wallace, *The Comanches*, 18; Smithwick, *Evolution of a State*, 136; John Salmon Ford, *Rip Ford's Texas by John Salmon Ford*, ed. Stephen B. Oats (Austin: University of Texas Press, 1963), 131; James Pike says that he saw a well-proportioned seven-foot-tall Comanche man on the San Gabriel River around 1859; James Pike, *The Scout and Ranger: Being the Personal Adventures Corporal Pike of the Fourth Ohio Cavalry* (Coppell, TX, 2020).
42. Newcomb, *Indians of Texas*, 90, 89.
43. In the early days, before overgrazing, grassfires swept the prairie clean, forming a distinct timber line. James Greer, ed., *James Buckner Barry, A Ranger and Frontiersman: The Days of Buck Barry in Texas, 1845–1906* (Dallas: Southwestern Press, 1932), n.p.
44. Wallace and Hoebel, *The Comanches*, 257.
45. Wallace and Hoebel, *The Comanches*, 259; Oats, *Rip Ford's Texas*, 135.
46. Oats, *Rip Ford's Texas*, 135.

47. Wallace and Hoebel, *The Comanches*, 259.
48. Sarah Catherine Lattimore, *Incidents in the History of Dublin by Sarah Catherine Lattimore*, ed. Mollie Louise Grisham (copied for the Dublin Public Library, 1967), 20–21. The crossed arms gesture of friendship is thought to have been popularized by the Masonic presence on the frontier; Kenneth Franklin Neighbours, *Robert Simpson Neighbors and the Texas Frontier, 1836–1859* (Waco: Texian Press, 1975), 44.
49. Jihong Cole-Dai, David Ferris, et al., "Paleoceanography and Paleoclimatology: Cold Decade (AD 1810–1819) Caused by Tambora (1815) and Another (1809) Stratospheric Eruption," *Advancing Earth and Space Science* (November 2009): abstract.
50. Llerena Friend, ed., *M. K. Kellogg's Texas Journal, 1872* (Austin: University of Texas Press, 1967), 132; Cole-Dai, Ferris, et al., *Cold Decade*.
51. William A. Klingaman and Nicholas P. Klingaman, *The Year without a Summer: 1816 and the Volcano That Darkened the World and Changed History* (New York: St. Martin's Griffin, 2013), 32–38.
52. Matthew D. Therrell and Makayla J. Trotter, "Waniyetu Wowapi: Native American Records of Weather and Climate," *American Meteorological Society* (May 2011): 583–92; the conjecture of a winter so disruptive that hide documents were lost is mine.
53. Klingaman and Klingaman, *Year without a Summer*, 201; Stefan Bronnimann and Daniel Kramer, "Tambora and the 'Year without a Summer' of 1816, a Perspective on Earth and Human Systems Science," *Geographica Bernensia* (2016): 42.
54. Noyes, *Los Comanches*, 197.
55. Anderson, *Conquest*, 47; Anderson, *Indian Southwest*, 145.
56. Susan L. Swan, "Drought and Mexico's Struggle for Independence," *Environmental Review* 6, no.1 (Spring 1982): 54–62; Luz Marina Arias and Luis de las Calle, "The Legacy of War Dynamics on State Capacity: Evidence from 19th Century Mexico," *Centro de Investigacion y Docencia Economomicas*, no. 612 (2018): 6.
57. Campbell, *Gone to Texas*, 96–97; Brian R. Hamnett, "Royalist Counterinsurgency and the Continuity of Rebellion: Guanajuato and Michoacán, 1813–20," *Hispanic American Historical Review* 62, no. 1 (1982): 19–48.
58. Anderson, *Indian Southwest*, 251; Hämäläinen, *Comanche Empire*, 190–91.
59. Hämäläinen, *Comanche Empire*, 191.
60. Smith, *From Dominance to Disappearance*, 122–23.
61. Gillen D'Arcy Wood, "The Volcano That Changed the Course of History," *Conversation* (2014): 2–3; Campbell, *Gone to Texas*, 101.
62. Campbell, *Gone to Texas*, 102–5.

63. David Hackett Fischer, *Albion's Seed: Four British Folkways in America* (New York: Oxford University Press, 1989), 760.
64. Fischer, *Albion's Seed*, 760, 618.
65. Anderson, *Conquest*, 7.
66. Smith, *Dominance to Disappearance*, 123.
67. Migrants originally from southern Britain are the Anglos; the Scots Irish originated in the borderlands between Scotland and England, and in Northern Ireland.
68. Anderson, *Indian Southwest*, 256.
69. Hämäläinen, *Comanche Empire*, 190; Anderson, *Indian Southwest*, 256.
70. David La Vere, *Life among the Texas Indians: The WPA Narratives* (College Station: Texas A&M University Press, 1998), 25–26; Hämäläinen, *Comanche Empire*, 193.
71. Anderson, *Indian Southwest*, 258–59.
72. Hämäläinen, *Comanche Empire*, 94.
73. Anderson, *Indian Southwest*, 263.
74. Anderson, *Indian Southwest*, 263.
75. Anderson, *Conquest*, 75; Ewers, *Indians of Texas*, 43.
76. Anderson, *Indian Southwest*, 264.
77. Anderson, *Conquest*, 75.
78. Robert Bernstein, "Public Health," *Texas Handbook Online* (June 15, 2010); Anderson, *Conquest*, 393n.
79. John F. Taylor, "Sociocultural Effects of Epidemics on the Northern Plains" (PhD diss., University of Montana, 1982), 54.
80. Stephen L. Moore, *Savage Frontier: Rangers, Riflemen, and Indian Wars in Texas 1, 1835–1837* (Denton: University of North Texas Press, 2002), 29; Robertson, *Rotting Face*, 176.
81. Robertson, *Rotting Face*, xii, xiii, 283–86; the epidemic was not spread by the Missouri River steamboat *St. Peter's*; Taylor, *Sociocultural Effects of Epidemics*, 56–57.
82. Anderson, *Conquest*, 100–101.
83. Noyes, *Los Comanches*, 225; La Vere, *Texas Indians*, 183; La Vere, *WPA Narratives*, 31.
84. O. T. Hayward et al., *A Field Guide to the Grand Prairie of Texas: Land, History, Culture* (Waco, TX: Baylor Program for Regional Studies, 1992), 47.
85. Moore, *Savage Frontier*, 131.
86. T. R. Fehrenbach, *Comanches: The Destruction of a People* (New York: Da Capo Press, 1974), 284–85; the white flag deception had been used the month before the Parker Raid; Anderson, *Conquest*, 128–29; Hayward et al., *Field Guide to the Grand Prairie*, 47.

87. Moore, *Savage Frontier*, 135–39; Treva Elaine Hodges, "The Captivity Narratives of Cynthia Ann Parker: Settler Colonialism, Collective Memory, and Cultural Trauma" (PhD diss., University of Louisville, 2019), 6–9.
88. Hodges, *Captivity Narratives*, 10; Anderson, *Conquest*, 408n; Moore, *Last Stand of the Cherokees*, 29; Robertson, *Rotting Face*, 56–57.
89. Hämäläinen, *Comanche Empire*, 215; Fehrenbach, *Destruction of a People*, 310, 314–15.
90. Fehrenbach, *Destruction of a People*, 314–15.
91. Moore, *Last Stand of the Cherokees*, 133–63.
92. Fehrenbach, *Destruction of a People*, 314–15.
93. Hämäläinen, *Comanche Empire*, 216.
94. Noyes, *Los Comanches*, 283.
95. Noyes, *Los Comanches*, 282; Smith, *Dominance to Disappearance*, 175.
96. Noyes, *Los Comanches*, 83.
97. Rena Maverick Green, ed., *Memoirs of Mary A. Maverick Arranged by Mary A. Maverick and Her Son George Madison Maverick* (San Antonio: Alamo Printing, 1921), 32.
98. Maverick Green, *Memoirs*, 38.
99. Noyes, *Los Comanches*, 284.
100. Noyes, *Los Comanches*, 285.
101. Smith, *Dominance to Destruction*, 176.
102. Noyes, *Los Comanches*, 176; Anderson, *Conquest*, 176.
103. Moore, *Savage Frontier*, 228; Smith, *Dominance to Disappearance*, 177.
104. Hämäläinen, *Comanche Empire*, 216–17.
105. Anderson, *Conquest*, 194.
106. Smith, *Dominance to Disappearance*, 177.
107. Anderson, *Conquest*, 189, 206; Moore, *Savage Frontier*, volume 3, 22–23; Heather Fishel, "The 1847 Colt Walker: The Most Powerful Handgun Ever Used by the U.S," *War History Online: The Place for Military History News and Views* (February 2018): 1; Bruce A. Glasrud and Harold J. Weiss Jr. *Tracking the Texas Rangers: The Nineteenth Century* (Denton: University of North Texas Press, 2012), 98.
108. Glasrud and Weiss, *Tracking the Texas Rangers*, 98.
109. Fishel, *1847 Walker Colt*, 1.
110. Anderson, *Conquest*, 207.
111. Bernstein, *Public Health*, 2; Wallace, *The Comanches*, 149; Taylor, *Sociocultural Effects of Epidemics*, 60.
112. Fishel, *1847 Walker Colt*, 1.
113. Fishel, *1847 Walker Colt*, 1.
114. Anderson, *Conquest*, 207.

115. Bernstein, "Public Health," 2.
116. Klingaman and Klingaman, *Year without a Summer*, 275.
117. Bernstein, "Public Health," 2.
118. Taylor, "Effects of Epidemics," 64.
119. Taylor, *Effects of Epidemics*, 66.
120. Neighbours, *Robert Simpson Neighbors*, 110.

Chapter Ten

1. Smithwick, *The Evolution of a State*, 127.
2. Moore, *Savage Frontier*, 126; H. Allen Anderson, "The Delaware and Shawnee Indians and the Republic of Texas, 1820–1845," *Southwestern Historical Quarterly* 94, no. 2 (October 1990): 240, 247.
3. Moore, *Savage Frontier*, volume 4, 126; Carol A. Lipscomb, "Delaware Indians," *Handbook of Texas History*, Texas State Historical Association; Anderson, "The Delaware and Shawnee Indians," 247.
4. Texas General Land Office, "Devoted to Peace: Delaware Chief John Conner," Texas: Save Texas History (November 2016); Lipscomb, "Delaware Indians," *Handbook of Texas*, Texas State Historical Association.
5. Anderson, "The Delaware and Shawnee," 239.
6. Anderson, "The Delaware and Shawnee," 242.
7. John Conner's Grant was near Haskell, Texas, General Land Office, "Devoted to Peace"; Lipscomb, *Delaware Indians*.
8. Texas General Land Office, "Devoted to Peace."
9. Neighbours, *Robert Simpson Neighbors*, 33; Anderson, *The Delaware and Shawnee*, 248.
10. Anderson, *The Delaware and Shawnee*, 248.
11. Rupert N. Richardson and H. Allen Anderson, "Shaw, Jim," *Handbook of Texas*, Texas State Historical Association.
12. Richardson and Anderson, *Shaw, Jim*.
13. Richardson, "Chisholm, Jesse," *Handbook of Texas*, Texas State Historical Association; Stan Hoig, *Jesse Chisholm: Ambassador of the Plains* (Niwor: University of Colorado Press, 1991).
14. Anderson, *Conquest*, 200.
15. Anderson, *Conquest*, 201; Moore, *Savage Frontier*, 126.
16. Becky Tanner, "150 Years Ago, Jesse Chisholm Opened Wichita's First Business," *Wichita Eagle*, January 2013.
17. Joseph McCoy, "Indian Cattle, Cattle Stealing," *Symphony in the Flint Hills: Field Journal* (2017): 17.
18. Anderson, "Delaware and Shawnee," 248, 249, 257, 262.

19. Anderson, "Delaware and Shawnee," 253.
20. Anderson, *Conquest*, 255, 335.
21. Ewers, *The Indians of Texas*, 52, 160.
22. Jim Robinson, "Wildcat's Roar—Hooah!," *Orlando Sentinel*, April 2001; Donald A. Swanson, "Coacoochee (Wildcat)," *Handbook of Texas*, Texas State Historical Association; the term *Seminole* was probably derived from the Spanish *cimarron*, escaped slave or wild. In the late sixteenth and early seventeenth centuries Muskogean Creeks and escaped Africans moved to Florida and settled on lands once occupied by the mostly extinct Natives. The term *maroon*, also a derivative of *cimarron*, referred to free and escaped Black people entering Florida. Many of these maroons lived in separate villages, carried arms, and had their own leaders They were more like vassals than slaves, a term only used later to defend the maroons from slave hunters. The two groups slowly merged over time; Kevin Mulroy, *The Seminole Freedmen: A History* (Norman: University of Oklahoma Press, 2016), 5–7.
23. Andrew K. Frank, "Creating a Seminole Enemy: Ethnic and Racial Diversity in the Conquest of Florida," *FIU Law Review* 9, no. 2 (Spring 2014): 277–79.
24. Robertson, "Wildcat's Roar—Hooah!"; Anderson, *Conquest*, 200.
25. Anderson, *Conquest*, 200–201; Swanson, "Coacoochee (Wildcat)."
26. Swanson, "Coacoochee (Wildcat)."
27. Kenneth W. Porter, "The Seminole in Mexico, 1850–1861," *Hispanic American Review* 31, no. 1 (February 1951): 1–2.
28. Wild Cat and John Horse traveled to Washington, DC, in an unsuccessful attempt to have the Seminoles recognized as a tribe and to put a stop to Creek raids; Philip Thomas Tucker, "John Horse: Forgotten African-American Leader of the Second Seminole War," *Journal of Negro History* 77, no. 2 (1992): 74–83; Porter, "Seminole in Mexico," 2; Neighbours, *Robert Simpson Neighbors*, 43.
29. Porter, "Seminole in Mexico," 5.
30. Kenneth W. Porter, "Davy Crockett and John Horse: A Possible Origin of the Coonskin Story," *American Literature* 15, no. 1 (March 1943): 10–15.
31. E. Douglas Sivad, "Caballo, Jaun," *Handbook of Texas*, Texas State Historical Association.
32. Anderson, *Conquest*, 200; Neighbours, *Robert Simson Neighbors*, 43.
33. William C. Yancy, "In Justice to Our Allies: The Government of Texas and Her Indian Allies, 1836–1867" (master's thesis, University of North Texas, 2008), 70.
34. Anderson, *Conquest*, 215.
35. Porter, "Seminole in Mexico," 5; William Loren Katz, *Black Indians: A Hidden Heritage* (New York: Simon and Schuster, 2012) 77.

36. Mulroy, *Seminole Freedmen*, 8; Porter, "Seminole in Mexico," 5; Anderson, *Conquest*, 238.
37. Porter, "Seminole in Mexico," 5.
38. Jon May, "Horse, John (ca. 1812–1882)," *Encyclopedia of Oklahoma History and Culture*, www.okhistory.org/publicatioins/enc./entry?entry=HO033. Anderson, *Conquest*, 238; Porter, "Seminole in Mexico," 6; many of their descendants, known as Indios Mascogos, still live there today; Mulroy, *Seminole Freedmen*, 8–9.
39. Mulroy, *Seminole Freedmen* 9.
40. Porter, "Seminole in Mexico," 7; Sean Kelly, "Mexico in His Head: Slavery and the Texas-Mexico Border, 1810–1860," *Journal of Social History* 37, no. 3 (Spring 2004): 718.
41. A letter by Robert S. Neighbors cited in Anderson, *Conquest*, 238.
42. Rupert N. Richardson, "Neighbors, Robert Simpson," *Handbook of Texas*.
43. Campbell, *Gone to Texas*, 202.
44. Newcomb, *Indians of Texas*, 354; Neighbours, *Robert Simpson Neighbors*, 152–60.
45. Campbell, *Gone to Texas*, 202; Neighbours, *Robert Simpson Neighbors*, 162; Smith, *Dominance to Disappearance*, 212.
46. Hämäläinen, *Comanche Empire*, 292, 303.
47. Hämäläinen, *Comanche Empire*, 307.
48. Neighbours, *Robert Simpson Neighbors*, 159.
49. La Vere, *The Texas Indians*, 198.
50. Campbell, *Gone to Texas*, 202–3.
51. Lucy A. Erath and her father, George B. Erath, *The Memoirs of Major George B. Erath, 1813–1891* (The Heritage Society of Waco, 1926), 87; Dan M. Young, "Stephenville, Texas," *Handbook of Texas*; Young, "Erath County," *Handbook of Texas*; Young, *1989 Historical Calendar of Erath County* (Stephenville, TX: Vanderbilt Street Press, 1989).
52. Young, *Historical Calendar, 1989*; H. G. Perry, *Grand Ol' Erath: The Saga of a Texas West Cross Timbers County* (Stephenville: Stephenville Printing, 1974), 4.
53. Young, *Historical Calendar, 1989*.
54. According to oral tradition, Red Jack is buried near the bridge over the Bosque River on Highway 377; Young, *Historical Calendar, 1989*; there is an alternative story that says as Red Jack was leaving the Stephenville square, he stopped at a cabin where a Black woman was giving him something to eat. Dr. W. W. McNeill, who had been following the Anadarko, showed up and demanded that Red Jack leave. McNeill said the Indian was belligerent so he snapped a cap at him, which caused Red Jack to pull his knife; then he was

shot by McNeill. Mrs. Fred H. Chandler, paper read before the Assembly of Club Ladies, Stephenville (April 18, 1913), 2.
55. Young, *Historical Calendar, 1989*; Oats, *Rip Ford's Texas*, 223.
56. Young, *Historical Calendar, 1987*; Neighbours, *Robert Simpson Neighbors*, 208; La Vere, *Texas Indians*, 199.
57. Oats, *Rip Ford's Texas*, 223–27; Neighbours, *Robert Simpson Neighbors*, 222.
58. Oats, *Rip Ford's Texas*, 229.
59. The Second US Cavalry commanded by Colonel Albert Sidney Johnson included Robert E. Lee and sixteen other officers who became generals during the Civil War. Oats, *Rip Ford's Texas*, 231–32; Neighbours, *Robert Simpson Neighbors*, 223; La Vere, *Texas Indians*, 199.
60. Neighbours, *Robert Simpson Neighbors*, 222; Barbara Throp Wilkins, "Peter Garland: The Controversial Captain—Hero or Villain?" *Granbury Magazine*, Fall 1986.
61. Smith, *From Dominance to Disappearance*, 232; Young, *Historical Calendar, 1980*; Sherry Knight, *Vigilantes to Verdicts: Stories from a Texas District Court*, (Stephenville: Jacobus Books, 2009) 39.
62. Neighbours, *Robert Simpson Neighbors*, 224–25; Anderson, *Conquest*, 314.
63. Young, *Historical Calendar, 1980*.
64. Samuel Stephen was the first person buried in Stephenville's West End Cemetery. Knight, *Vigilantes*, 40; Neighbours, *Robert Simpson Neighbors*, 224; Wilkins, "Peter Garland"; Young, *Historical Calendar, 1980*; Anderson, *Conquest*, 314.
65. Wilkins, "Peter Garland."
66. Knight, *Vigilantes*, 40–41; Fehrenbach, *Lone Star*, 504–5.
67. Neighbours, *Robert Simpson Neighbors*, 227; Fehrenbach, *Lone Star*, 504–5; Young, *Historical Calendar, 1980*.
68. Smith, *Dominance to Disappearance*, 234–35; Erath, *Memoirs of George B. Erath*, 90; Knight, *Vigilantes*, 40.
69. Smith, *Dominance to Disappearance*, 234; Erath, *Memoirs of George B. Erath*, 90.
70. Neighbours, *Robert Simpson Neighbors*, 227.
71. Knight, *Vigilantes*, 40.
72. Neighbours, *Robert Simpson Neighbors*, 226.
73. Oats, *Rip Ford's Texas*, xxxiii, 252; Smith, *Dominance to Disappearance*, 235.
74. Anderson, *Conquest*, 238.
75. Anderson, *Conquest*, 239; Jerry Don Thompson, "Colonel John Robert Baylor: Texas Indian Fighter and Confederate Soldier," A Hill Junior College Monograph, no. 5 (1971), 100.
76. Neighbours, *Robert Simpson Neighbors*, 209.
77. Fehrenbach, *Lone Star*, 503.

78. Noah Marshall Jr., "Reservation War," *Handbook of Texas*; Campbell, *Gone to Texas*, 203; Anderson, *Conquest*, 328.
79. Anderson, *Conquest*, 329; Neighbours, *Robert Simpson Neighbors*, 208; La Vere, *Texas Indians*, 199.
80. Neighbours, *Robert Simpson Neighbors*, 208; Anderson, *Conquest*, 329.
81. Anderson, *Conquest*, 330.
82. Fehrenbach, *Lone Star*, 505; La Vere, *Texas Indians*, 200.
83. Neighbours, *Robert Simpson Neighbors*, 239.
84. Camp Cooper was an army post founded by Colonel Albert Sidney Johnson in 1856 on the Clear Fork of the Brazos seven miles above Fort Griffin, to monitor the nearby Comanche reservation. There were soldiers within the camp who cooperated with Garland to help him capture the artillery and deliver it to Baylor. Something went wrong. Charles G. Davis, "Camp Cooper," *Handbook of Texas*; Neighbours, *Robert Simpson Neighbors*, 239.
85. Neighbours, *Robert Simpson Neighbors*, 239.
86. Smith, *Dominance to Disappearance*, 239.
87. The William Marlin family was on friendly terms with the reservation Indians and agents. Zacariah E. Coombes and Barbara Neal Ledbetter, eds., *The Diary of a Frontiersman, 1858–1859* (Newcastle, TX: Barbara Neal Ledbetter, 1962), 32, 35; Neighbours, *Robert Simpson Neighbors*, 239; Smith, *Dominance to Disappearance*, 239.
88. Neighbours, *Robert Simpson Neighbors*, 244.
89. Neighbours, *Robert Simpson Neighbors*, 244–45.
90. Neighbours, *Robert Simpson Neighbors*, 245.
91. The Alabama-Coushatta were allowed to stay in Texas on their East Texas reservation; the day of departure was kept secret because of Baylor's threats to attack the columns. Smith, *Dominance to Disappearance*, 242.
92. Smith, *Dominance to Disappearance*, 242; Neighbours, *Robert Simpson Neighbors*, 261.
93. Neighbours, *Robert Simpson Neighbors*, 271.
94. Neighbours, *Robert Simpson Neighbors*, 278.
95. Neighbours, *Robert Simpson Neighbors*, 279.
96. Neighbours, *Robert Simpson Neighbors*, 283.
97. Smith, *Dominance to Disappearance*, 244.
98. Anderson, *Conquest*, 330–31; Judith Ann Benner, *Sul Ross: Soldier, Statesman, Educator* (College Station: Texas A&M University Press, 1983), 47–51.
99. Glen Sample Ely, "Myth, Memory, and Massacre: The Pease River Capture of Cynthia Ann Parker (review)," *Southwestern Historical Quarterly* 115, no. 1 (July 2010).

100. Paul H. Carlson, *Myth, Folklore, and Misconception in the 1860 Capture of Cynthia Ann Parker*," in *Legends and Life in Texas*, ed. Kenneth Untiedt (Denton: University of North Texas Press, 2017), 67.
101. Carlson, *Myth, Folklore, and Misconception*, 67.
102. Benner, *Sul Ross*, 54.
103. Ely, "Myth, Memory, and Massacre"; Carlson, *Myth, Folklore, and Misconception*, 71; Benner, *Sul Ross*, 177.
104. Benner, *Sul Ross*, 55; Sul Ross corrected this boast in 1893. Carlson, *Myth Folklore, and Misconception*, 66–68.
105. Carlson, *Myth, Folklore, and Misconception*, 67; Wilbarger, *Indian Depredations in Texas: Reliable Accounts* (Austin: Hutchings Printing House, 1889), 338.
106. La Vere, *The Texas Indians*, 202–3; Richard Seager and Celine Herweijer, "Causes and Consequence of Nineteenth Century Droughts in North America," *Lamont-Doherty Earth Observatory of Columbia University: Drought Research*, 2.
107. Anderson, *Conquest*, 327.
108. Anderson, *Conquest*, 330.
109. La Vere, *The Texas Indians*, 203.
110. Seager and Herweijer, *Causes and Consequence*, 4.
111. Hämäläinen, *Comanche Empire*, 298.
112. Fehrenbach, *Comanches*, 316.
113. Hämäläinen, *Comanche Empire*, 313–15.
114. The Western Cross Timbers and Prairie pockets were burned nearly every year during the Civil War drought, which was beneficial for the grasses and cleared woody growth in preparation for a quick reestablishment. Michael C. Stambaugh, Richard P. Guyette, et al., "Fire, Drought, and Human History near the Western Terminus of the Cross Timbers, Wichita Mountains, Oklahoma, USA," *Fire Ecology* 5 (2009): 51–65; Hämäläinen, *Comanche Empire*, 313.
115. Hämäläinen, 212; Anderson, *Conquest*, 347.
116. Anderson, *Conquest*, 347.
117. Hämäläinen, *Comanche Empire*, 313–14.
118. Hämäläinen, *Comanche Empire*, 320, 324; Anderson, *Conquest*, 347.
119. Hämäläinen, *Comanche Empire*, 325; Anderson, *Conquest*, 346–47; Fehrenbach, *Lone Star*, 392.
120. Hämäläinen, *Comanche Empire*, 331.
121. Hämäläinen, *Comanche Empire*, 332.
122. Campbell, *Gone to Texas*, 291; Fehrenbach, *Comanches*, 495; Hämäläinen, *Comanche Empire*, 330.
123. Fehrenbach, *Comanches*, 486–87.

124. Hämäläinen, *Comanche Empire*, 328; Campbell, *Gone to Texas*, 490.
125. La Vere, *The Texas Indians*, 271.
126. Michael D. Pierce, *The Most Promising Young Officer: A Life of Ranald Slidell Mackenzie* (Norman: University of Oklahoma Press, 1993), 111–16.
127. Campbell, *Gone to Texas*, 293; Pierce, *Most Promising Young Officer*, 123.
128. Pierce, *Most Promising Young Officer*, 142.
129. Fehrenbach, *Comanches*, 513.
130. La Vere, *The Texas Indians*, 212; Hämäläinen, *Comanche Empire*, 36.
131. La Vere, *The Texas Indians*, 212.
132. Weston La Barre, *The Ghost Dance: The Origins of Religion* (Maidstone, UK: Crescent Moon Publishing, 1970), 170, 282.
133. Hämäläinen, *Comanche Empire*, 337; Gaines Kinkaid, "Isa-Tai," *The Handbook of Texas*.
134. La Barre, *Ghost Dance*, 282.
135. The Comanche version of the Plains Sun Dance did not involve tortuous breast piercings or dragging around bison skulls from thongs passed through back muscles. It was not a vision-seeking experience. Earnest Wallace and E. Adamson Hoebel, *The Comanches: Lords of the South Plains* (Norman: University of Oklahoma Press, 1952), 320–24.
136. Wallace and Hoebel, *The Comanches*, 325.
137. Wallace and Hoebel, *The Comanches*, 325; Gaines Kincaid, *Isa-Tai*; Hämäläinen, *Comanche Empire*, 337.
138. Wallace and Hoebel, *The Comanches*, 327; Pierce, *Most Promising*, 149–61.
139. Wallace and Hoebel, *The Comanches*, 327.
140. Noyes, *Los Comanches*, 309.
141. Pierce, *Most Promising*, 169.
142. Zesch, *The Captured*, 219.
143. Zesch, *The Captured*, 219; Herman Lehmann was captured and adopted by Apaches at eleven; later he joined the Comanches and fought against Texas Rangers, the US Cavalry, and raided Anglo settlers. Stephen Taylor, "Alvin C. Greene, Jr.," *Handbook of Texas*.
144. Zesch, *The Captured*, 223.
145. Paul Roland, "Home on the Range: The Impact of the Cattle Trails on Indian Territory," *Armstrong Undergraduate Journal of History* 8, no. 2, Article 4 (2018): 58.
146. Loretta Fowler, *The Columbia Guide to American Indians of the Great Plains* (New York: Columbia University Press, 2003), 177; La Vere, *The Texas Indians*, 216–17; Roland, *Home on the Range*, 60.
147. Roland, *Home on the Range*, 60–64.
148. Perry, *Grand Ol' Erath*, 22.

149. Anderson, *Conquest*, 330; Young, *Historical Calendar, 1987*; Anderson, *Conquest*, 330.
150. An article by Joe Fitzgerald published in a Stephenville newspaper in 1925 said that the Jenny Papworth murder at McDow Hole after the Civil War was committed by Anglos disguised as Indians. Young, *Historical Calendar, 1981*.
151. Mary Joe Clendenin, *The Ghost of the McDow Hole* (New York: Carlton Press, 1979), 18–19; Thomas T. Ewell, *History of Hood County: From Its Earliest Settlement to the Present* (Granbury, TX: Granbury News, 1895), 32.
152. Young, *1987 Historical Calendar*.
153. Young, *1989 Historical Calendar*.
154. Young, *1980 Historical Calendar*.
155. Young, *1989 Historical Calendar*; Young, *1979–1989 Historical Calendars*.
156. While researching for the historical calendars, I interviewed residents of several nursing homes in the area. One thing that stuck with me was that the people who lived on the small farms, though poor, looked back on their lives as hard, but satisfying. The selling out and moving to town was tragic because they lost their tight-knit communities. Young, *1979–1989 Historical Calendars*.

Bibliography

Abbott, Peter Michael, Ulrike Niemeier, Claudia Timmreck, F. Riede, J. R. McConnell, M. Severi, and Hubertus Fischer. "Volcanic Climate Forcing Preceding the Inception of the Younger Dryas: Implication for Tracing the Laacher See Eruption." *Quaternary Science Reviews* 274 (2021): 107260.

Acuna-Soto, Rodolfo, David W. Stahle, Malcolm K. Cleaveland, and Matthew D. Therrell. "Megadrought and Megadeath in 16th Century Mexico." *Emerging Infectious Diseases* 8, no. 4 (2002): 360–62.

Adair, Mary J., Neil A. Duncan, Danielle N. Young, Steven R. Bozarth, and Robert K. Lusteck. "Early Maize (*Zea mays*) in the North American Central Plains: The Microbotanical Evidence." *American Antiquity* 87, no. 2 (2022): 333–51.

Ajivsgi, Greyata. *Wildflowers of Texas*. Bryan, TX: Shearer Publishing, 1984.

Anderson, Gary Clayton. *The Indian Southwest, 1580–1830: Ethnogenesis and Reinvention*. Norman: University of Oklahoma Press, 1999.

Anderson, H. Allen. "The Delaware and Shawnee Indians and the Republic of Texas, 1820–1845." *Southwestern Historical Quarterly* 94, no. 2 (October 1990): 231–60.

Ansley, R. J., J. A. Huddle, and B. A. Kramp. "Mesquite Ecology." *Texas Natural Resources Server* (1997). httpa://texnat.tamu.edy/library/symposia/brush-sculptors . . . /mesquite-ecology/.

Arias, Luz Marina, and Luis de las Calle. "The Legacy of War Dynamics on State Capacity: Evidence from 19th Century Mexico." *Centro de Investigacion y Docencia Economomicas*, no. 612 (2018).

Axsmith, Brian J., and Bonnie Fine Jacobs. "The Coniffer Frenelopsis Ramosissima (Cheirolepidiaceae) in the Lower Cretaceous of Texas: Systematic, Biogeographical, and Paleoecological Implications." *International Journal of Plant Science* 166 (2005): 327–37.

Ayala, Sergio J. "Calf Creek Horizon Evidence at the Gault Site (41BL323): A Description of the Imagery Found in the Volume 5 Cover Border Design." *Index of Texas Archeology: Open Access Gray Literature from the Lone Star State*, no. 44 (2019): xi.

Bader, Jurgen, Johann Jungclaus, Natalie Krivova, Stephan Lorenz, Amanda Maycock, Thomas Raddatz, Hauke Schmidt, Matthew Toohey, Chi-Ju Wu, and Martin Claussen. "Global Temperature Modes Shed Light on the Holocene Temperature Conundrum." *Nature Communications* 11, no. 1 (2020): 4726.

Barbieri, R., M. Signoli, D. Cheve, C. Costedoat, S. Tzortzis, G. Aboudharam, D. Raoult, and M. Drancourt. "Yersinia Pestis: The Natural History of Plague." *Clinical Microbiology Reviews* 34, no. 1 (2020): 10–28.

Bard, Edouard, and Martin Frank. "Climate Change and Solar Variability: What's New Under the Sun?" *Earth and Planetary Science Letters* 248 (2006): 1–14.

Barlow, Connie. *The Ghost of Evolution: Nonsensical Fruit, Missing Partners, and Other Ecological Anachronisms*. New York: Basic Books, 2000.

Barr, Juliana. "There Is No Such Thing as 'Prehistory': What the Longue Duree of Caddo and Pueblo History Tells Us about Colonial America." *William and Mary Quarterly* 74, no. 2 (April 2017): 203–40.

Baskin, John. "The Pleistocene Fauna of South Texas." *Kingsville: Texas A&M University*. users.tamuk.edu/kfjab02/SOTXFAUN.htm.

Baugh, Timothy G., and Jay C. Blaine. "Enduring the Violence: Four Centuries of Kirikir'i's Warfare." *Plains Anthropologist* 62, no. 242 (2017): 99–132.

Bell, Willis Harvey, and Edward Franklin Castetter. *The Utilization of Yucca, Sotol, and Beargrass by the Aborigines in the American Southwest*. Albuquerque: University of New Mexico, Ethnobotanical Studies in the Southwest, 1941.

Bement, Leland C. *Hunter-Gatherer Mortuary Practices during the Central Texas Archaic*. Austin: University of Texas Press, 1994.

Benner, Judith Ann. *Sul Ross: Soldier, Statesman, Educator*. College Station: Texas A&M University Press, 1983.

Bennion, Michael Kay. *Captivity, Adoption, Marriage, and Identity: Native American Children in Mormon Homes, 1847–1900*. Las Vegas: UNLV Theses, Dissertations, Professional Papers, and Capstones, 2012.

Benson, Larry V., Michael S. Berry, Edward A. Jolie, Jerry D. Spangler, David W. Stahle, and Eugenen M. Hattori. "Possible Impacts of Early-11th-, Middle-12th-, and Late-13th-Century Droughts on Western Native Americans and the Mississippian Cahokians." *Quaternary Science Reviews* 26, no. 3–4 (2007): 336–50.

Berbesque, J. Colette, and Kara C. Hoover. "Frequency and Developmental Timing of Linear Enamel Hypoplasia Defects in Earl Archaic Texan Hunter-Gatherers." *PeerJ* (2018): e4367.

Bernstein, Robert. "Public Health." *Texas Handbook Online* (June 15, 2010). https://www.tshaonline.org/handbook/entries/public-health.

Beschta, Robert, L., and William J. Ripple. "Yellowstone's Prehistoric Bison: A Comment of Keigley (1019)." *Rangelands* 41, no. 3 (2019): 149–51.

Bishop, Arthur L. "Flood Potential of the Bosque Basin." *Baylor Geological Studies*, no. 33 (1977).

Black, Steve. "Spiro and the Arkansas Basin, Caddo Fundamental." *Texas Beyond History*, The University of Texas (August 6, 2003).

Black, Steve, and Vaughn M. Bryant Jr. "Hinds Cave: A Perishable Scientific Treasure: Life at Hinds." *Texas Beyond History*, The University of Texas (2005). www.texasbeyondhistory.net/hinds/.

Black, Steve, and Emily McCuistion. *Life in the Bosque River Basin and Beyond*. Austin: University of Texas Press, 2022.

Blitz, John H. "Adoption of the Bow in Prehistoric North America." *North American Archeologist* 9, no. 2 (1988): 123–37.

Boldurian, Anthony T., and Susanne M. Hubinsky. "Preforms in Folsom Lithic Technology: A View from Blackwater Draw, New Mexico." *Plains Anthropologist* 39, no. 150 (1994): 445–64.

Booth, Robert K., Stephen T. Jackson, Steven L. Forman, John E. Kutzbach, E. A. Bettis III, Joseph Kreigs, and David K. Wright. "A Severe Centennial-Scale Drought in Midcontinental North America 4200 Years Ago and Apparent Global Linkages." *The Holocene* 15, no. 3 (2005): 321–28.

Booth, Robert K., John E. Kutzbach, Dara C. Hotchkiss, and Reid A. Bryson. "A Reanalysis of the Relationship between Strong Westerlies and Precipitation in the Great Plains and Midwest Regions of North America." *Climate Change* 76 (June 2006): 427–41.

Bourgeon, Lauriane, and Ariane Burke. "Horse Exploitation by Beringian Hunters during the Last Glacial Maximum." *Quaternary Science Reviews* 269 (2021): 107140.

Bousman, C. Britt, Michael B. Collins, Paul Goldberg, Thomas Stafford, Jan Guy, Barry W. Baker, and D. Gentry Steele. "The Palaeoindian-Archaic Transition in North America: New Evidence from Texas." *Antiquity* 76, no. 294 (2002): 980–90.

Boyd, Douglas Kevin, John E. Dockall, Karl W. Kibler, Gemma Mehalchick, Laura M. Short, and Chester P. Walker. *Data Recovery Investigations at the Tank Destroyer Site (41CV1378) at Fort Hood, Coryell County, Texas*. Austin: Prewitt and Associates, 2014.

Bradley, Raymond S., and Jostein Bakke. "Is There Evidence for a 4.2ka BP Event in the Northern North Atlantic Region?" *Climates of the Past* 15 (2019): 1665–76.

Bronaugh, Whit. "The Trees That Miss the Mammoths." *American Forests* (2015), www.americanforests.org/magazine/article/trees-that-miss-the-mammals/.

Bronniman, Stefan, and Daniel Kramer. "Tambora and the 'Year without a Summer' of 1816: A Perspective on Earth and Human Systems Science." *Geographica Bernensia* 90 (2016).

Brown, Ken. "Archeomalacology: What We Can Learn from Snails." *The TARL Blog: The Texas Archeological Research Laboratory, Texas Beyond History*, 2015.

Bryant, Vaughn M., Jr., and Richard G. Holloway. "A Late-Quaternary Paleoenvironmental Record of Texas: An Overview of the Pollen Evidence." In *Pollen Records of Late-Quaternary North American Sediments*, 1985, edited by Vaughn M. Bryant Jr. and Richard G. Holloway, 39–70. American Association of Stratigraphic Palynologists Foundation, 1985.

Buchanan, Brian Andres, Michael J. O'Brien, and Metin I. Eren. "An Assessment of Stone Weapon Tip Standardization during the Clovis-Folsom Transition in the Western United States." *American Antiquity* 83, no. 4 (2018): 721–34.

Buchanan, Briggs, Brian Andrews, J. David Kilby, and Metin I. Eren. "Settling into the Country: Comparison of Clovis and Folsom Lithic Networks in Western North America Shows Increasing Redundancy of Toolstone Use." *Journal of Anthropological Archaeology* 53 (2019): 32–42.

Buchanan, Briggs, J. David Kilby, Jason M. LaBelle, Todd A. Surovell, Jacob Holland-Lulewicz, and Marcus J. Hamilton. "Bayesian Modeling of the Clovis and Folsom Radiocarbon Records Indicates a 200-Year Multigenerational Transition." *American Antiquity* 87, no. 3 (2022): 567–80.

Budd, Jon, Tim Perttula, John Dockall, and Karl W. Kibler. *Testing and Data Recovery Excavations at the Jayroe Site (41HM51), Hamilton County, Texas (Waco District, CSJ No. 0909-29-030)*. Austin: Prewitt and Associates, Cultural Resources of Texas, Report 187, I-100, I-123 (2020).

Buntgen, Ulf. "Global Wood Anatomical Perspective on the Onset of the Late Antique Little Ice Age (LALIA) in the Mid-6th Century." *Science Bulletin* 67, no. 22 (2022): 2336–44.

Buntgen, Ulf, Dominique Arseneault, Etienne Boucher, Olga V. Churakova, Fabio Gennaretti, Alan Crivellaro, and Malcolm K. Hughes. "Prominent Role of Volcanism in Common Era Climate Variability and Human History." *Dendrochronologies* 64 (2020): 125757.

Buntgen, Ulf, Vladimir S. Myglan, Fredrik Charpentier Ljungqvist, Michael McCormick, Nicola Di Cosmo, Michael Sigl, and Johann Jungclaus. "Cooling and Societal Change during the Late Antique Little Ice Age from 536 to around 660 AD." *Nature Geoscience* 9, no. 3 (2016): 231–36.

Bush, Leslie L. "Evidence for a Long-Distance Trade in Bois d'arc Bows in 16th Century Texas (*Maclura pomifera* Moraceae)." *Journal of Texas Archeology and History* 1 (2014): 51–69.

Bush, Mark B. *Ecology of a Changing Planet*. San Francisco: Benjamin Cummings, 2020.

Bustos, David, and Thomas M. Urban. "Reply to 'Evidence for Humans at White Sands National Park during the Last Glacial Maximum Could Actually Be before Clovis People—13,000 Years Ago' by C. Vance Hayes, Jr." *PaleoAmerica: A Journal of Early Migration and Dispersal* 8, no. 2 (2022): 99–101.

Campbell, Randolph B. *Gone to Texas: A History of the Lone Star State*. New York: Oxford University Press, 2003.

Carlson, David L., Barry W. Baker, William A. Dickens, John E. Dockall, and Lee C. Nordt. *Archeological Investigations along Owl Creek: Results of the 1992 Summer Archaeological Field School*. Fort Hood, TX: United States Army Fort Hood, Archeological Resource Management Series, Research Report, Number 21, 1997.

Carlson, Gustav, and Volney Jones. "Some Notes on Uses of Plants by the Comanche Indians." *Papers of the Michigan Academy of Science, Arts, and Letters* 25 (1930): 5217–542.

Carlson, Kristen, and Leland Bements. "Bison across the Holocene: What Did Calf Creek Foragers Hunt? In *The Calf Creek Horizon*, edited by John C. Lohse, Marjorie A. Duncan, and Don G. Wyckoff, 16–23. College Station: Texas A & M University Press, 2021.

Carlson, Paul H. "Myth Folklore and Misconception in the 1860 Capture of Cynthia Ann Parker." In *Legends and Life in Texas: Folklore from the Lone Star State, in Stories and Song*, edited by Kenneth L. Untiedt. Publication of the Texas Folklore Society LXXI. Denton: University of North Texas Press, 2017.

Carpenter, Stephen M. "Long-Term Subsistence Strategies from the Archaic to Late Prehistoric Times." In *The Siren Site and the Long Transition from Archaic to Late Prehistoric Lifeways on the Eastern Edwards Plateau of Central Texas*, edited by Stephen M. Carpenter, Kevin A. Miller, Mary Jo Galindo, Brett Houk, and Charles D. Frederick. *Index of Texas Archeology: Open Access Gray Literature from the Lone Star State*, 2013, Article 4.

Cary, Jennifer H. "*Quercus Marilandica*: In: Fire Effects Information System." US Department of Agriculture, Forest Service, Rocky Mountain Research Station, Fire Sciences Laboratory. 2019. www.fs.fed.us/database/feis/plants/tree/quemar/all.html.

Castro, Joseph. "First Americans Used Spear-Throwers to Hunt Large Animals." *Live Science*. Last update January 28, 2015. www.livescience.com/49603-paleoindian-spear-thrower-evidence.html.

Chandler, Mrs. Fred H. Paper read before the Assembly of Club Ladies, Stephenville, Texas, April 18, 1913.

Chatters, James C., Douglas J. Kennett, Yemane Asmerom, Brian M. Kemp, Victor Polyak, Ablerto Nava Blank, and Patricia A. Beddos. "Late Pleistocene Human Skeleton and mtDNA Link Paleoamericans and Modern Native Americans." *Science* 344, no. 6185 (2014): 750–54.

Chew, Sing C. *The Recurring Dark Ages: Ecological Stress, Climate Changes, and System Transformation*. New York: Altamira Press, 2007.

Clark, John W. "Implications of Land and Fresh-Water Gastropods in Archeological Sites." *Journal of the Arkansas Academy of Science* 23, no. 9 (1969): 38–54.

Clark, Jorie, Anders E. Carlson, Albero V. Reyes, Elizabeth C. B. Carlson, Louise Guillaume, Glenn A. Milne, Lev Tarasov, Marc Caffee, Klaus Wilcken, and Dylan H. Rood. "The Age of the Opening of the Ice-Free Corridor and Implication for the Peopling of the Americas." *Proceedings of the National Academy of Sciences* 119, no. 14 (2022): e2118558119.

Clarke, Mary Whatley. *Chief Bowles and the Texas Cherokees*. Norman: University of Oklahoma Press, 1971.

Clendenin, Mary Joe. *The Ghost of the McDow Hole*. New York: Carlton Press, 1979.

"Clovis Reconsidered: The Gault Site." www.texasbeyondhistory.net/gault/clovis.html.

Cobb, Kim M., Christopher D. Charles, Hai Cheng, R. and Lawrence Edwards. "El Niño/Southern Oscillation and Tropical Pacific Climate during the Last Millennium." *Nature* 424 (July 2003): 271–76.

Collins, Michael B., and C. Britt Bousman. "Cultural Implications of Late Quaternary Environmental Change in Northeastern Texas." *CRHR Research Reports* 1 (2015): 1–80.

Connell, J. H., director. *The Peach*. College Station Texas Agricultural Experiment Station, no. 39 (July 1896).

Coombes, Zacariah E., and Barbara Neal Ledbetter, ed. *The Diary of a Frontiersman, 1858–1859*. Newcastle, TX: Barbara Neal Ledbetter, 1962.

Cordova, Carlos E., and William C. Johnson. "An 18ka to Present Pollen- and Phytolith-Based Vegetation Reconstruction from Hall's Cave, South-Central Texas, USA." *Quaternary Research* 92, no. 2 (2019): 497–518.

Crook, Wilson W., III. *The Carrollton Phase Archaic: A Redefinition of the Chronology, Composition, and Aerial Distribution of the Early Archaic Horizon along the Trinity River, Texas*. Houston: Houston Archeological Society, Report No. 35, 2020.

Crook, Wilson W., III, and Mark D. Hughston. "The Late Prehistoric of the East Fork of the Trinity River." *CRHR Research Reports* 2, no. 1 (2016): 1.

Dalquest, Walter W. "Mammals of the Pleistocene Slaton Local Fauna of Texas." *Southwestern Naturalist* 12, no. 1 (April 20, 1967): 1–30.

Dampier, Bennett Harrison. "The Moth and the Moonflower: Datura and Hawk Moth Iconography across Ancient America." Master's thesis, Texas State University at San Marcos, 2022.

Daniels, J. Michael, and James C. Knox. "Alluvial Stratigraphic Evidence for Channel Incision during the Mediaeval Warm Period on the Central Great Plains, USA." *Holocene* 15, no. 5 (2005). https://doi.org/10.1191/0959683605hl847rp.

Danielson, Dennis, and Karl Reinhard. "Human Dental Microwear Caused by Calcium Oxalate Phytoliths in Prehistoric Diet of the Lower Pecos Region, Texas." *American Journal of Anthropology* 107 (1998): 297–304.

Davis, Charles G. "Camp Cooper." *Handbook of Texas Online*. https://www.tshaonline.org/handbook/entries/camp-cooper.

Davis, Dan R., Jr. *Prehistoric Artifacts of the Texas Indians: An Identification and Reference Guide*. Fort Sumner, NM: Pecos Publishing 1991.

Davis, Kaitlyn Elizabeth. "The Ambassador's Herb: Tobacco Pipes as Evidence for Plains-Pueblo Interaction, Interethnic Negotiation, and Ceremonial Exchange in the Northern Rio Grande." Master's thesis, University of Colorado at Boulder, 2017.

Davis, Loren G., and David B. Madsen. "The Coastal Migration Theory: Formulation of Testable Hypotheses." *Quaternary Science Reviews* 249 (2020): 106605.

Dean, Bradie. "Caddo Artifacts in Central Texas: A Proposed Trade Connection." Master's thesis, Baylor University, 2020.

Deeds, Susan M. "Colonial Chihuahua: Peoples and Frontiers in Flux." In *New Views of Borderland History*, edited by Robert H. Jackson. Albuquerque: University of New Mexico Press, 1998.

De Melo, Fernando Lucas, Joana Carvalho Moreira De Mello, Ana Maria Fraga, Kelly Nunes, and Sabine Eggers. "Syphilis at the Crossroad of Phylogenetics and Paleopathology." *PLoS Neglected Tropical Diseases* 4, no. 1 (2010): e575.

De Pastino, Blake. "16,000-Year-Old Tools Discovered in Texas, among the Oldest Found in the West." *Western Digs*. Last updated July 18, 2016. http://westerndigs.org/16,000-year-old-tools-in-texas-among-oldest-yet-found-in-the-west/15.

Dering, Phil. "Daily Bread and Healing Balm: A Deep History of Native Plant Use in the Trans-Pecos of Texas." *The Sabal*, 23, no. 1 (2006).

Dial, Susan, and Albert Redder. "A Paleoindidan Grave: The Horn Shelter." The University of Texas at Austin. Last updated December 2010. https://www.texasbeyondhistory.net/horn/burials.html.

Dimitrov, Theodor. "In the Times of Madness and Fury: Observations on Collective Behavioral Deviations in Byzantium in the Age of Justinian Plague (541–750)." *Central and Eastern European Online Library*, 27 (2021).

Dobie, J. Frank. *The Longhorns*. Austin: University of Texas Press, 1982.

Dobson, Jerome E., Giorgio Spada, and Gaia Galassi. "The Bering Transitory Archipelago: Stepping Stones for the First Americans." *Geoscience* 353, no. 1 (2021): 55–65.

Dockall, John E., Ross C. Fields, Karl W. Kibler, Cory J. Broehm, Jon Budd, Eloise F. Gadus, and Karen M. Gardner. *Testing and Data Recovery Excavations at the Jayroe Site (41HM51), Hamilton County, Texas (Waco District, CSJ No. 0909-29-030)*. Austin: Texas Department of Transportation, 2020.

Dong, Guanghui, M. Wei, Y. Yang, R. Liu, J. Wang, L. Chen, and M. Lu. "A Brief History of Wheat Utilization in China." *Front Agr. Sci. Eng* 6 (2019): 288–95.

Dorsey, George A. *The Mythology of the Wichita*. Washington, DC: Carnegie Institution of Washington, DC, 1904.

Douglas, Frederick H. "The Wichita Indians and Allied Tribes: Waco, Towakoni, and Kichi." *Denver Art Museum* 40 (January 1932): 2.

Dull, Robert A., John R. Southon, Steffen Kutterolf, Kevin J. Anchukaitis, Armin Freundt, David B. Wahl, and Payson Sheets. "Radiocarbon and Geologic Evidence Reveal Ilopanga Volcano as Source of the Colossal 'Mystery' Eruption of 539/40 CE." *Quaternary Science Reviews* 222 (2019): 105855.

Dycus, Katy. "After Cooper's Ferry, Rethinking How the Americas Were Peopled." *Mammoth Trumpet: Center for the Study of the First Americans* 3, no. 2 (2023): 8–20.

Elias, Scott Armstrong. "First Americans Lived on Bering Land Bridge for Thousands of Years." *The Conversation*. February 28 2014, 6:21 a.m. EST. https://theconversation.com.Fossil

Elias, Scott Armstrong. "Late Pleistocene Climates of Beringia, Based on Analysis of Fossil Beetles." *Quaternary Research* 53, no. 2 (2017): 229–32.

Ely, Glen Sample. "Myth, Memory, and Massacre: The Pease River Capture of Cynthia Ann Parker (review.)" *Southwestern Historical Quarterly* 115, no. 1 (July 2010): 91–92.

Englar, Mary. *The Comanches: Nomads of the Southern Plains*. Mankato, MN: Capstone Press, 2004.

Enoch, Mary-Ann, and Bernard J. Albaugh. "Genetic and Environmental Risk Factors for Alcohol Use Disorders in American Indians and Alaskan Natives." *America Journal of Addictions* (August 2017): 461–68.

Erath, Lucy A., and George B. Erath. *The Memoirs of Major George B. Erath, 1813–1891*. Waco, TX: Heritage Society of Waco, 1926.

Eren, Metin I., David J. Meltzer, Brett Story, Briggs Buchanan, Don Yeager, and Michell R. Bebber. "On the Efficacy of Clovis Fluted Points for Hunting Proboscideans." *Journal of Archaeological Science: Reports* 390 (2021): 103166

Eren, Metin I., Michelle R. Bebber, Edward J. Knell, Brett Story, and Briggs Buchanan. "Plains Paleoindian Projectile Point Penetration Potential." *Journal of Anthropological Research* 78, no. 1. (2022): 84–112.

Erickson, John R., and Douglas K. Boyd. *Porch Talk: A Conversation about Archeology in the Texas Panhandle*. Lubbock: Texas Tech University Press, 2022.

Everett, Dianna. *The Texas Cherokees: A People between Two Fires, 1819–1840*. Norman: University of Oklahoma Press, 1990.

Ewell, Thomas T. *History of Hood County: From Its Earliest Settlement to the Present*. Granbury, TX: Granbury News, 1895.

Ewers, John C., ed. *The Indians of Texas in 1830 by Jean Luis Berlandier*. Washington, DC: Smithsonian Institution Press, 1969.

Fehrenbach, T. R. *Comanches: The Destruction of a People*. New York: Da Capo Press, 1994.

———. *Lone Star: A History of Texas and the Texans*. New York: Wingbooks, 1968.

Fenn, Elizabeth A. *Pox Americana: The Great Smallpox Epidemic of 1775–82*. New York: Hill and Wang, 2001.

Fields, Ross C. "The Prairie Caddo Model and the J. B. White Site." *Index of Texas Archeology: Open Access Gray Literature from the Lone Star State 2017*, no. 1 (2017): 43.

Firmin, Mark Edward. "For the Pleasure of the People: A Centennial History of William Cameron Park, Waco, Texas." PhD thesis, Baylor University, 2009.

Fischer, David Hackett. *Albions's Seed: Four British Folkways in America*. New York: Oxford University Press, 1989.

Fischer, Kristine. "Form and Function: A Case Study Using Pedernales Points from the Gault Site (41BL323) in Central Texas," Master's thesis, University of Exeter, 2015.

Fishel, Heather. "The 1847 Colt Walker: The Most Powerful Handgun Ever Used by the U.S." *War History Online: The Place for Military History New and Views* (2018). https://www.war history.com.

Flint, M. L., ed. "Wild Blackberries." Agriculture and Natural Resources, University of California, 2019.

Flores, Dan. *American Serengeti: The Last Big Animals of the Great Plains*. Lawrence: University of Kansas Press, 2016.

Flores, Dan, ed. *Journal of an Indian Trader: Anthony Glass and the Texas Trading Frontier, 1790–1810*. College Station: Texas A&M University Press, 1985.

Ford, John Salmon. *Rip Ford's Texas*, edited by Stephen B. Oates. Austin: University of Texas Press, 1987.

Foster, William C. *Climate and Culture Change in North America, AD 900–1600*. Austin: University of Texas Press, 2012.

Fowell, Sarah, and David Scholl. "The Bering Strait, Rapid Climate Change, and Land Bridge Paleoecology." Final report of the JOI/USSSP/IARC Workshop, Fairbanks, Alaska, June 20–22, 2005.

Fowler, Loretta. *The Columbia Guide to American Indians of the Great Plains*. New York: Columbia University Press, 2003.

Fowles, Severin. "The Pueblo Village in the Age of Reformation (AD 1300–1600)." In *The Oxford Handbook of North American Archeology*, edited by Timothy R. Pauketat, 631–44. Oxford: Oxford University Press, 2012.

Frank, Andrew K. "Creating a Seminole Enemy: Ethnic and Racial Diversity in the Conquest of Florida." *FIU Law Review* 9, no. 2 (Spring 2014): 277–93.

Friend, Llerena, ed. *M. K. Kellogg's Texas Journal, 1872*. Austin: University of Texas Press, 1967.

Froese, Duane, Mathias Stiller, Peter D. Heintzman, Alberto V. Reyes, Grant D. Zazula, Andre E. R. Soares, and Matthias Meyer. "Fossil and Genomic Evidence Constrains the Timing of Bison Arrival in North America." *Proceedings of the National Academy of Sciences* 119, no. 13 (2017): 3457–62.

Gaetani, Marco, Gabriele Messori, M. Carmen Alvarez Castro, Qiong Zhang, and Francesco S. R. Pausata. "Mid-Holocene Climate at Mid-Latitudes: Modelling the Impact of the Green Sahara." *EGU General Assembly Conference Abstracts* (2022): EGU22–2031

Galindo, Mary Jo, Kevin A. Miller, and Stephen M. Carpenter. "Metric Discrimination of Projectile Points from 41WM1126." In *The Siren Site and the Long Transition from Archaic to Late Prehistoric Lifeways on the Eastern Edwards Plateau of Central Texas*, edited by Stephen M. Carpenter, Kevin A. Miller, Mary Jo Galindo, Brett Houk, and Charles D. Frederick. Austin: Index of Texas Archeology: Open Access Gray Literature from the Lone Star State, 2013, Article 4.

Galloway, B. T. (Chief of Bureau). *Agricultural Varieties of the Cowpea and Immediately Related Species*. Washington, DC: US Department of Agriculture, no. 229.

Gayton, Anna Hardwick. "The Narcotic Plant Datura in Aboriginal American Culture." PhD dissertation, University of California at Berkeley, 1926.

Gibbon, Guy. "Lifeways through Time in the Upper Mississippi River Valley and Northeastern Plains." In *The Oxford Handbook of North American Archeology*, edited by Timothy Pauketat, 325–35. Oxford: Oxford University Press, 2012.

Gleim, Elizabeth, R., L. Mike Conner, Roy D. Berghaus, Michael L. Levin, Galina E. Zemtsova and Michael J. Yabsley. "The Phenology of Ticks and the Effects of Long-Term Prescribed Burning on Tick Population Dynamics in Southwestern Georgia and Northwestern Florida." *PLoS One* 9, no. 11. (2014). https://doi.org/10.1371/journal.pone.0112174.

Gould, Frank W. *The Grasses of Texas*. College Station: Texas A&M University Press, 1975.

Grauke, L. J. "Hickory." In *Nut Tree Culture in North America*, edited by Dennis Fulbright, 117–66. State University of New York, Northern Nut Growers Association, 2003.

Grauke, L. J. "Geographic Patterns of Genetic Variation in Native Pecans." *Tree Genetics and Genomes* 7, no. 5 (2011): 917–32.

Green, Monica, Lori Jones, Lester K. Little, Uli Schamiloglu, and George D. Sussman. "Yersinia Pestis and the Three Plague Pandemics." *The Lancet Infectious Diseases* 14, no. 10 (2014): 918.

Green, Rena Maverick. ed. *Memoirs of Mary A. Maverick Arranged by Mary A. Maverick and Her Son George Madison Maverick*. San Antonio: Alamo Printing, 1921.

Greer, James, ed. *James Buckner Barry, a Ranger and Frontiersman: The Days of Buck Barry in Texas, 1845–1906*. Dallas: Southwestern Press, 1932.

Grund, Brigid Sky. "Behavioral Ecology, Technology, and the Organization of Labor: How a Shift from Spear Thrower to Self Bow Exacerbates Social Disparities." *American Anthropologist* 119, no. 1 (2017): 104–19.

Hai, Chewng, Haiwei Zhangm, Chrisoph Spotl, and R. Lawrence Edwards. "Timing and Structure of the Younger Dryas Event and Its Underlying Climate Dynamics," edited by Mark Thiemens. *Proceedings of the National Academy of Sciences* 117, no. 38 (2020): 1–10.

Hall, Douglas W. "Hydrologic Significance of Depositional Systems and Facies in Lower Cretaceous Sandstones, North-Central Texas." *Geological Circular* 76, no. 1 (1976): 9–10.

Hall, Grant D. *Allen's Creek: A Study in the Cultural Prehistory of the Lower Brazos River Valley, Texas*. Austin: Texas Archeological Survey Research Report, 61, University of Texas Press, 1981.

Hall, Stephen A., Thomas W. Boutton, Christopher R. Lintz, and Timothy G. Baugh. "New Correlation of Stable Carbon Isotopes with Changing Late-Holocene Fluvial Environments in the Trinity River Basin of Texas, USA." *The Holocene* 22, no. 5 (2012): 541–49.

Hämääainen, Pekka. *The Comanche Empire*. New Haven, CT: Yale University Press, 2008.

———. *Indigenous Continent: The Epic Contest for North America*. New York: Liveright Publishing, 2022.

Hamilton, Danny L. *Prehistory of the Rustler Hills Granado Cave*. Austin: University of Texas Press, 2001.

Hamnett, Brian R. "Royalist Counterinsurgency and the Continuity of Rebellion: Guanajuato and Michoacan, 1813–20." *Hispanic American Historical Review* 62, no. 1 (1982): 19–48.

Harris, Susan Meriwether. "The Western Cross Timbers: Scenario of the Past, Outcome for the Future." Master's thesis, Texas Christian University, 2008.

Hart, John P. "Evolving the Three Sisters: The Changing Histories of Maize, Bean, and Squash in New York and the Greater Northeast." In *Current Northeast Paleoethnobotany II*, edited by John P. Hart. New York State Museum Bulletin No. 494, University of the State of New York, State Education Department, 2008.

Hart, John P. "Maize Agriculture Evolution in the Eastern Woodlands of North America: A Darwinian Perspective." *Journal of Archeological Method and Theory* 6, no. 2 (1999): 137–80.

Hart, John P., and William A. Lovis. "Reevaluating What We Know about the Histories of Maize in Northeastern North America: A Review of Current Evidence." *Journal of Archeological Research* 21 (2013): 175–216.

Hass, Randall, James Watson, Tammy Buonasera, John Southon, Jennifer C. Chen, Sarah Noe, Kevin Smith, Carlos Viviano Llave, Jelmer Eerkens, and Gendon Parker. "Female Hunters of the Early Americas." *Science Advances* 6, no. 45 (2020): eabd0310.

Hatch, Peter. "We Abound in the Luxury of the Peach." *Twinleaf* 10 (1998): 9–11.

Havard, V. "Food Plants of the North American Indians." *Bulletin of the Torrey Botanical Club* 22, no. 22 (1985): 98–123.

Hayward, O. T., P. N. Dolliver, D. L. Amsbury, and J. C. Welderman. *A Field Guide to the Grand Prairie of Texas: Land, History and Culture*. Baylor Program for Regional Studies, 1992.

Heinrich, Bernd. *Mind of the Raven: Investigations and Adventures with Wolf Birds*. New York: HarperCollins, 2009.

Herrman, Edward W., and G. William Monaghan. "Post-Glacial Drainage Basin Evolution in the Midcontinent, North America: Implications for Prehistoric Human Settlement Patterns." *Quaternary International* 511 (2019): 68–77.

Hixson, Charles, and Buddy Whitley. "Early and Late Toyah-Period Occupations within Area C of the Baker Site on the Northeastern Edwards Plateau, Central Texas." *Journal of Texas Archeology and History* 7, Article 2 (April 13, 2023).

Hixson, Chuck. "Graham-Applegate Rancheria: What Is the Austin Phase (and the Late Prehistoric)?" *Texas Beyond History* (2001).

Hoig, Stan. "Chisholm, Jesse." *The Encyclopedia of Oklahoma History and Culture* (January 15, 2010). https://ww.okhistory.org/publications/enc/entry?entry=CH067.

Hodges, Glenn. "First Americans." *National Geographic Magazine*, January 2015, 1–11.

Hodges, Treva Elaine. "The Captivity Narratives of Cynthia Ann Parker: Settler Colonialism, Collective Memory, and Cultural Trauma." PhD dissertation, University of Louisville, 2019.

"Honey Locust" Plant Fact Sheet, *United States Department of Agriculture Natural Resources Conservation Services*.

"Horn Shelter: Archaic Fishhook Manufacturing." *Texas Beyond History*, texas beyondhistory. net/horn/manufacturing.html.

Hudson, Charles M., ed. *Black Drink: A Native American Tea*. Athens: University of Georgia Press, 1979.

Hunziker, Johanna. "Exploring Burned Rock Middens at Camp Bowie." *Texas Beyond History*, Center for Archeological Research, University of Texas at San Antonio. April 18, 2004.

"Ice Age (Pleistocene Epoch)." Gulf of Mexico Program. Last updated 11/2/2009. http://www.epa.gov/gmpo/edresources/pleistocene.html.

Javier. "Impact of the 2,400 Year Solar Cycle on Climate and Human Societies." *Climate, Etc*. Posted on September 20 2016. https://judithcurry.com/2016/09/20/impact-of-the-2,400-yr-solar-cycle-on-climate-and-human-societies.

Jelks, Edward B. *The Kyle Site: A Stratified Central Texas Aspect Site in Hill County, Texas*. Austin: University of Texas Press #5, 1962.

Jennings, Thomas A., and Ashley M. Smallwood. "Clovis and Toyah: Convergent Blade Technologies on the Southern Plains Periphery of North America." In *Convergent Evolution in Stone-Tool Technology*, edited by Michael J. O'Brien, Briggs Buchanan, and Metin I. Eren, 229–53. Cambridge, MA: MIT Press, 2018.

Jihong, Cole Dai, David A. Ferris, Alyson Lanciki, Joel Savarino, and Melanie Baroni. "Cold Decade (AD 1810–1819) Caused by Tambora (1815) and Another (1809) Stratospheric Volcanic Eruption." *Geophysical Research Letters* 36, no. 22 (2009): L22703.

John, Elizabeth A. H. *Storms Brewed in Other Men's Worlds: The Confrontation of Indians, Spanish, and French in the Southwest, 1540–1795*. Norman: University of Oklahoma Press, 1996.

Johnson, Eileen, Stance Hurst, and John A. Moretti. "Late Quaternary Stratigraphy and Geochronology of the Spring Creek Drainage along the Southern High Plains Eastern Escarpment, Northwest Texas." *Quaternary* 4, no. 3 (2021): 1–37.

Johnson, LeRoy. *The Life and Times of Toyah-Culture Folk.* Austin: Texas Department of Transportation and Texas Historical Commission, Office of the State Archeologist Report, 38 (1994).

Jones, Karen. "The Story of Comanche: Horsepower, Heroism, and the Conquest of the American West." *War and Society* 36, no. 3 (2017): 156–81.

Katz, William Loren. *Black Indians: A Hidden Heritage.* New York: Simon and Schuster, 2012.

Kaufman, Darrell, Nicholas McKay, Cody Routson, Michael Erb, Christoph Datwyler, Philipp S. Sommer, Oliver Heiri, and Basil Davis. "Holocene Global Mean Surface Temperature, a Multi-Method Reconstruction Approach." *Scientific Data* 7, no. 1 (2020): 201.

Kelly, Sean. "Mexico in His Head: Slavery and the Texas-Mexico Border, 1810–1860." *Journal of Social History* 37, no. 3 (Spring 2004): 709–23.

Kibler, Karl W., and Tim Gibbs. "Archeological Survey of 61 Acres along the Bosque River, Waco, McLennan County, Texas." Ross C. Fields, Principal Investigator, *Prewitt and Associates, Inc., Technical Reports* 69 (July 2004).

Kibler, Karl W., and Gemma Mehalchick. *Hunters and Gatherers of the North Bosque River Valley: Excavations at the Baylor, Britton, McMillan, and Higginbotham Sites, Waco Lake, McMillan County, Texas.* Austin: Prewitt and Associates, July 2008.

———. "Hunter-Gatherer Resource Acquisition and Use in the Lower Bosque River Basin during the Late Archaic." *Bulletin of the Texas Archeological Society* 81 (2010): 103–26.

Kilham, Lawrence. *The American Crow and the Common Raven.* College Station: Texas A&M University Press, 1989.

Kindscher, Kelly. *Edible Wild Plants of the Prairie: An Ethnobotanical Guide.* Lawrence: University Press of Kansas, 1987.

———. *Medicinal Wild Plants of the Prairie: An Ethnobotanical Guide.* Lawrence: University Press of Kansas, 1992.

King, Megan M., Roger Cain, and Shawna Morton Cain. "An Experimental Ethnoarchaeological Approach to Understanding the Development of Use Wear Associated with the Processing of River Cane for Split-Cane Technology." *Southeastern Archeology* 38, no. 1 (2018): 38–53.

Kincaid, Gaines. "Isa-Tai," *Handbook of Texas.*

Klingaman, William A., and Nicholas P. Klingaman. *The Year without a Summer: 1816 and the Volcano That Darkened the World and Changed History.* New York: St. Martin's Griffin, 2013.

Knight, Sherry. *Vigilantes to Verdicts: Stories from a Texas District Court.* Stephenville: Jacobus Books, 2009.

Koenig, Walter D., and Johannes M. H. Knops. "The Mystery of Masting in Trees." *American Scientist* 93 (2005): 340–47.

Kozuch, Laura. "Olivella Beads from Spiro and the Plains." *American Antiquity* 67, no. 4 (2002): 697–709.

Kvernes, Kimberly K., Marie E. Blake, Karl W. Kibler, Jennifer K. McWilliams, E. Frances Gadus, and Ross C. Fields. "Relocation and Updated Recordation of 44 Archeological Sites at Waco Lake." US Army 20000628 (2000): 31.

La Barre, Weston. *The Ghost Dance: The Origins of Religion*. Maidstone, UK: Crescent Moon Publishing, 1970.

———. *The Peyote Cult*. Norman: University of Oklahoma Press, 2012.

Lapointe, Francois, and Raymond S. Bradley. "Little Ice Age Abruptly Triggered by Intrusion of Atlantic Waters in the Nordic Seas." *Science Advances* 7, no. 51 (2021): eabi8230.

Larsen, Clark Spencer. *Skeletons in Our Closet: Revealing Our Past through Bioarchaeology*. Princeton, NJ: Princeton University Press 2000.

Larson, Donald A., Vaughn M. Bryant, and Tom S. Patty. "Pollen Analysis of a Central Texas Bog." *American Midland Naturalist* 88 (1972): 358–67.

Lattimore, Sarah Catherine. *Incidents in the History of Dublin*, edited by Mollie Louise Grisham. Copied for the Dublin, Texas, Public Library, 1967.

La Vere, David. *Life among the Texas Indians: The WPA Narratives*. College Station: Texas A&M University Press, 1998.

———. *The Texas Indians*. College Station: Texas A&M University Press, 2004.

Lawrence, Gregory B., James W. Sutherland, Charles W. Boylen, S. W. Nierzwicki-Bauer, Bahram Momen, Barry P. Baldigo, and Howard A. Simonin. "Acid Rain Effects on Aluminum Mobilization Clarified by Inclusion of Strong Organic Acids." *Environmental Science & Technology* 41, no. 1 (2007): 93–98.

Legare, J. D., ed. *The Southern Agriculturalist, Vol. III, 1830*. Charleston: A. E. Miller, 1830.

Lemke, Ashley K., D. Clark Wernecke, and Michael Collins. "Early Art in North America from the Clovis and Later Paleoindian Incised Artifacts from the Gault Site, Texas (41BL323)." *American Antiquity* 80, no. 1 (2015): 113–33.

Leonhardy, Frank C. *Domebo: A Paleo-Indian Mammoth Kill in the Prairie-Plains*. Great Plains Historical Society, 1966.

Lewand, Raymond L., Jr. "The Geomorphic Evolution of the Leon River System." *Baylor Geological Studies* 17 (1969): 5–27.

Leydet, David J., Anders E. Carlson, James T. Teller, Andrew Breckenridge, Aaron M. Barth, David J. Ullman, Gaylen Sinclair, Glenn A. Milne, Joshua K. Cuzzone, and Marc W. Caffee. "Opening of Glacial Lake Agassiz's Eastern Outlets by the Start of the Younger Dryas Cold Period." *Geology* 46, no. 2 (2018): 155–58.

"Life in the Bosque River Basin and Beyond." *Texas Beyond History*.

Lintz, Christopher. "Texas Panhandle-Pueblo Interactions from the Thirteenth Century through the Sixteenth Century." In *Farmers, Hunters, and Colonists: Interaction between the Southwest and the Southern Plains*, edited by Kathrine A. Spielmann, chapter 6. Tucson: University of Arizona Press, 1991.

Lintz, Christopher, Stephen A. Hall, Timothy G. Baugh, and Tiffany Osburn. "Archeological Testing at 41TR170, along the Clear Fork of the Trinity River, Tarrant County, Texas." *Texas Department of Transportation, Environmental Affairs Division*, Number 348 (2008).

Lipscomb, Carol A. "Delaware Indians." *Handbook of Texas History*. https://www.tshaonline.org/handbook/entries/delaware-indians.

Logan, Brad. "Late Woodland Feasting and Social Networks in the Lower Missouri River Region." *North American Archeologist* (2022): 1–46.

Lohse, Jon, Stephen L. Black, and Lalu M. Cholak. "Toward an Improved Archaic Radiocarbon Chronology for Central Texas." *Bull. Texas Archeol. Soc* 85 (2014): 259–87.

Lohse, Jon C., David B. Madsen, Rendon J. Culleton, and Douglas J. Kennett. "Isotope Paleoecology of Episodic Mid-to-Late Holocene Bison Population Expansions in the Southern Plains, USA." *Quaternary Science Reviews* 102 (2014): 14–26.

Lohse, Jon C., Don G. Wyckoff, and Marjorie Duncan. "The Calf Creek Horizon: Mid-Holocene Adaptions in North America." In *The Calf Creek Horizon*, edited by John C. Lohse, Marjorie A. Duncan, and Don G. Wyckoff, 3–15. College Station: Texas A&M University Press, 2021.

Magne, Martin P. R., and R. G. Matson. "Moving On: Expanding Perspectives on Athapaskan Migration." *Canadian Journal of Archaeology/Journal Canadien d'Archeologie* (2010): 212–39.

Majure, Lucas C., Raul Puente, M. Patrick Griffith, Walter S. Judd, Pamela S. Soltis, and Douglas E. Soltis. "Phylogeny of *Opuntia* ss (Cactaceae): Clade Delineation, Geographic Origins, and Reticulate Evolution." *American Journal of Botany* 99, no. 5 (2012): 847–64.

Malof, Andrew. "Feast or Famine: The Dietary Role of Rabdotus Species Snails in Central Texas." Master's thesis, University of Texas at San Antonio, 2001.

Mann, Barbara Alice. *The Tainted Gift: The Disease Method of Frontier Expansion*. Santa Barbara, CA: Praeger, 2009.

Marcy, Randolf B. Exploration of the Red River of Louisiana in the Year 1852. *War Department*, (1854). http://name.umdl.umich.edu/ABB2532.0001.001.

Marques, Andrea Horvath, Thomas G. O'Connor, Christine Roth, Ezra Susser, and Anne-Lise Bjorke-Monsen. "The Influence of Maternal Prenatal and Early Childhood Nutrition and Maternal Prenatal Stress on Offspring Immune

System Development and Neurodevelopmental Disorders." *Frontiers in Neuroscience* 7 (2013): 120.

Marshal, Noah F., Jr. "Reservation War." *Handbook of Texas Online*. www.tshaonline.org/handbook/entries/reservation-war.

Matero, I. S. O., and L. J. Gregoire. "The 8.2ka Cooling Event Caused by Laurentide Ice Saddle Collapse." *Earth and Planetary Science Letters* 473 (2017): 205–14.

Mathena, Sarah. "Developing a Multistage Model for Treponemal Disease Susceptibility." Master's thesis, Mississippi State University, 2013.

May, Jon. "Horse, John (ca. 1812–1882)." Encyclopedia of Oklahoma History and Culture. www.okhistory.org/publicatioins/enc./entry?entry=HO033.

McCoy, Joseph. "Indian Cattle/Cattle Stealing/Cattle." *Symphony in the Flint Hills Field Journal*, 2017. https://newprairie press.org/sfh/2017/fence/3.

McElrath, Dale L., Thomas E. Emerson, and Andrew C. Fortier. "Social Evolution or Social Response? A Fresh Look at the 'Good Gray Culture' After Four Decades of Midwest Research." In *Late Woodland Societies: Tradition and Transformation across the Mid-Continent*, edited by Thomas E. Emerson, Dale L. McElrath, and Andrew C. Fortier, 3–36. Lincoln: University of Nebraska Press, 2008.

McPherson, James K., and Gerald L. Thompson. "Competitive and Alleopathic Suppression of Understory by Oklahoma Oak Forests." *Bulletin of the Torrey Botanical Club* 99, no. 6 (1972): 293–300.

McWeeney, Lucinda. "Revising the Paleoindian Environmental Picture in Northeastern North America." In *Foragers of the Terminal Pleistocene in North America*, edited by Renee B. Walker and Boyce N. Driskell, Essay 9. Lincoln: University of Nebraska Press, 2007.

Meier, Holly A., Lee C. Nordt, Steven L. Forman, and Steven G. Driese. "Late Quaternary Alluvial History of the Middle Owl Creek Drainage Basin in Central Texas: A Record of Geomorphic Response to Environmental Change." *Quaternary International* 306 (2013): 24–41.

Meltzer, David J. "Human Responses to Middle Holocene (Altithermal) Climates on the North American Plains." *Quaternary Research* 52 (1999): 404–16.

"Microscopic Diamonds Suggest Cosmic Impact Responsible for Major Period of Climate Change." *Geology Times* 1 (2014).

Mildeo, Lauren. "Fieldwork Revises Ice-free Corridor Hypothesis of Human Migration." *Earth: The Science Behind the Headlines*. Last updated April 13, 2014. https://www.Earthmagazine.org/article/fieldwork-revises-ice-free-corridor-hypothesis-human-migration.

Moerman, Daniel E. *Native American Medicinal Plants: An Ethnobotanical Dictionary*. Portland, OR: Timber Press, 2009.

Mole Rain and Other Natural Phenomena in the Welsh Annals. https://www.academia.edu/MOLE-RAINAND-OTHER-NATURAL-PHENOMENA.

Monaghan, G. William, Timothy M. Schilling, and Kathryn E. Parker. "The Age and Distribution of Domesticated Beans (Phaseolus Vulgaris) in Eastern North America: Implications for Agricultural Practices and Group Interactions." *Occasional Papers* 1 (2014): 161–74.

Moore, Christopher R., Allen West, Malcolm A. LeCompte, Mark J. Brooks, I. Randolph Daniel Jr., Albert C. Goodyear, and Terry A. Ferguson. "Widespread Platinum Anomaly Documented at the Younger Dryas Onset in North American Sedimentary Sequences." *Scientific Reports* 7, no. 1 (2017): 44031.

Moore, Stephen L. *Last Stand of the Cherokees: Chief Bowles and the 1839 Cherokee War in Texas*. Garland, TX: RAM Books, 2009.

———. *Savage Frontier: Rangers, Riflemen, and Indian Wars in Texas, Volume IV, 1842–1845*. Denton: University of North Texas Press, 2010.

Moore, William. "A Guide to Texas Arrowpoints." Brazos Valley Research Associates, *Contribution in Archeology*, no. 6 (2015): 51.

Mueller, Natalie G. "The Occurrence of a Newly Described Domesticate in Eastern North America: Adena/Hopewell Communities and Agricultural Innovation." *Journal of Anthropological Archeology* 49 (2018).

———. "Seeds as Artifacts of Communities of Practice: The Domestication of Knotweed in Eastern North America." PhD dissertation, Washington University in Saint Louis, 2017.

Mueller, Natalie G., Gayle J. Fritz, Paul Patton, Stephen Carmody, and Elizabeth T. Horton. "Growing the Lost Crops of Eastern North America's Original Agricultural System." *Nature Plants* 3, no. 7 (2017): 1–5

Mulroy, Kevin. *The Seminole Freedmen: A History*. Norman: University of Oklahoma Press, 2016.

Munteanu, Nina. "The Power of Myth in Storytelling: The Alien Next Door." (2017). https://ninamunteanu.me/2017/03/28/the-power-of-myth-in-storytelling-2/.

"Mussel Collecting," *Texas Beyond History*.

Nabhan, Gary Paul. Native Seeds/Search Catalogue. nativeseeds.org.

Nasirpour, Mohammad Hossein, Abbas Sharifi, Mohsen Ahmadi, and Saeid Jafarzadeh Ghoushchi. "Revealing the Relationship between Solar Activity and COVID-19 and Forecasting of Possible Future Viruses Using Multi-Step Autoregression (MSAR)." *Environmental Science and Pollution Research* 28 (2012): 38074–84.

Navia, Carlos E. "On the Occurrence of Historical Pandemics during the Grand Solar Minima." *European Journal of Applied Sciences* 2, no. 4 (2020): 1–8.

Neighbours, Kenneth Franklin. *Robert Simpson Neighbors and the Texas Frontier, 1836–1859*. Waco: Texian Press, 1975.

Newcomb, W. W., Jr. *The Indians of Texas: From Prehistoric to Modern Times*. Austin: University of Texas Press, 1961.

Newton, Cody. "Towards a Context for Late Precontact Culture Change: Comanche Movement Prior to Eighteenth Century Spanish Documentation." *Plains Anthropologist* 56, no. 217 (2011): 53–69.

Nichols, Kerry. "Late Woodland Cultural Adaptions in the Lower Missouri River Valley: Archer, Warfare, and the Rise of Complexity." PhD dissertation, University of Missouri–Columbia, 2015.

Nordt, Lee C. "Archeological Geology of the Fort Hood Military Reservation Fort Hood, Texas." *United States Army, Fort Hood, Archeological Resource Management Series, Report 25*, (1992).

Norris, S. L., D. Garcia-Castellanos, J. D. Jansen, P. A. Carling, Martin Margold, R. J. Woysitka, and D. G. Froese. "Catastrophic Drainage from the Northwestern Outlet of Glacial Lake Agassiz during the Younger Dryas." *Geophysical Research Letters* 48, no. 15 (2021): e2021GL093919.

Noyes, Stanley. *Los Comanches: The Horse People, 1751–1845*. Albuquerque: University of New Mexico Press, 1993.

O'Brien, Michael J. "Setting the Stage: The Late Pleistocene Colonization of North America." *Quaternary* 2, no. 1 (2019): doi:10.3390/quat2010001.

Odezulu, Christopher I., Travis Swanson, and John B. Anderson. "Holocene Progradation and Retrogradation of the Central Texas Coast Regulated by Alongshore and Cross-Shore Sediment Flux Variability." *Depositional Record* 7, no. 1 (2021): 77–92.

Oppenheimer, Clive. *Eruptions That Shook the World*. Cambridge: Cambridge University Press, 2011.

Pappas, Stephanie. "Ancient Poop Gives Clues to Modern Diabetes Epidemic." *Live Science*, July 24, 2012.

Peng, Qian, Ian R. Gizer, Kirk C. C. Wilhelmsen, and Cindy L. Ehlers. "Associations between Genomic Variants in Alcohol Dehydrogenase Genes and Alcohol Symptomatology in American Indians and European Americans: Distinctions and Convergence." *Alcoholism: Clinical and Experimental Research* 41, no. 10 (2017): 1695–704.

Peppers, Krista Clements. "Old Growth Forests in the Western Cross Timbers of Texas." PhD dissertation, University of Arkansas, 2004.

Peppers, Krista Clements, and Richard V. Francaviglia. *The Cast Iron Forest: A Natural and Cultural History of the North American Cross Timbers*. Austin: University of Texas Press, 2000.

Peregrine, Peter N. "Climate and Social Change at the Start of the Late Antique Little Ice Age." *The Holocene* 30, no. 11 (2020): 1643–48.

Perkins, Stephen M. "Protohistory and the Wichita." *Plains Anthropologist* 53, no. 208 (November 2008): 381–94.

Perkins, Stephen M., Susan C. Vehik, and Richard R. Drass. "The Hide Trade and Wichita Social Organization: An Assessment of Ethnological Hypotheses Concerning Polygyny." *Plains Anthropologist* 53, no. 208 (2008): 432–43.

Perry, H. G. *Grand Ol' Erath: The Saga of a Texas West Cross Timbers County.* Stephenville, TX: Stephenville Printing, 1974.

Perttula, Timothy K. "Prairie Caddo Sites in Coryell and McLennan Counties in Central Texas." *Index of Texas Archaeology: Open Access Gray Literature for the Lone Star State 2016*, no. 1 (2016): 102.

Perttula, Timothy K., Duncan McKinnon, and Scott Hammerstedt. "The Archaeology, Bioarchaeology, Ethnography, Ethnohistory, and History Bibliography of the Caddo Indian Peoples of Arkansas, Louisiana, Oklahoma and Texas." *Index of Texas Archaeology: Open Access Gray Literature for Lone Star State 2021, Article 1* (2021).

Perttula, Timothy K., Robert Z. Selden Jr., and Jon C. Lohse. "Cultural Setting of the Leon River Basin." In *Archaeological and Geological Test Excavations at Site 41HM61, Hamilton County, Texas,* edited by Robert A. Weinstein, 15–27. Austin: Texas Department of Transportation, Environmental Affairs Division, Archeological Studies Program, 2015.

Peyton, Abigail, Stephen M. Carpenter, John D. Lowe, and Ken Lawrence. *Data Recovery Investigations on the Eastern Side of the Siren Site (42WM1126), Williamson County, Texas.* Austin: SWCA Environmental Consultants, December 2013.

Pierce, Michael D. *The Most Promising Young Officer: A Life of Ranald Slidell Mackenzie.* Norman: University of Oklahoma Press, 1993.

Pike, James. *The Scout and the Ranger: The Personal Adventures of Corporal Pike of the Fourth Ohio Calvary.* Coppell, TX, 2020.

"Plio-Pleistocene." American Museum of Natural History. https://research.amnh.org/palentology

Pluckhahn, Thomas J., Neill J. Wallis, and Victor D. Thompson. "The History and Future of Migrationist Explanations in the Archaeology of the Eastern Woodlands with a Synthetic Model of Woodland Period Migrations on the Gulf Coast." *Journal of Archaeological Research* 28, no. 4 (2020): 443–502.

Pollan, Michael. *The Botany of Desire: A Plants-Eye View of the World.* New York: Random House, 2001.

Ponkratova, Irina Y., Loren J. Davis, Daniel W. Bean, David B. Madsen, Alexander J. Nyers, and Ian Buvit. "Technical Similarities between-13ka

Stemmed Points from Ushki V, Kamchatka, Russian Far East, and the Earliest Stemmed Points in North America." In *Maritime Prehistory of Northeast Asia*, edited by Jim Cassidy, Irina Ponkratova, and Ben Fitzhugh, 233–61. Singapore: Springer, 2022.

Pope, Saxton T. *Bows and Arrows*. Berkeley: University of California Press, 1962.

Popejoy, Traci, Charles R. Randkiev, and Steve Wolverton. "Conservation Implications of Late Holocene Freshwater Mussels Remains of the Leon River in Central Texas." *Hydrobiologia* (2016): 1–8.

Porter, Joshua John. "Reconstructing Bison and Mammoth Migration during the Late Pleistocene and Early Holocene of Central Texas Using Strontium Isotopes." Master's thesis, University of Arkansas, 2022.

Porter, Kenneth W. "Davy Crockett and John Horse: A Possible Origin of the Coonskin Story." *American Literature* 15, no. 1 (March 1943): 10–15.

———. "The Seminole in Mexico, 1850–1861." *Hispanic American Review* 31, no. 1 (February 1951): 1–36.

Powell, James Lawrence. "Premature Rejection in Science: The Case of the Younger Dryas Impact Hypothesis." *Science Progress* 105, no. 1 (2022): 00368504211064272.

Pratt, Jordan, Ted Goebel, Kelly Graf, and Masami Izuho. "A Circum-Pacific Perspective on the Origin of Stemmed Points in North America." *PaleoAmerica* 6, no. 1 (2020): 64–108.

Prewitt, Elton R. "Bell: A Calf Creek Series Dart Point Type in Texas." In *The Calf Creek Horizon*, edited by John C. Lohse, Marjorie A. Duncan, and Don G. Wyckoff, 134–57. College Station: Texas A&M University Press, 2021.

Prikryl, Paul, Vojto Rusin, Emil A. Prikryl, Pavel Stastny, Maros Turna, and Martina Zelenakova. "Heavy Rainfall, Floods, and Flash Floods Influenced by High-Speed Solar Wind Coupling to the Magnetosphere—Ionosphere—Atmosphere System." *Annales Geophysicae* 39, no. 4 (2021): 769–93.

Quigg, J. Michael. Benjamin G. Bury, Robert A. Ricklis, Paul M. Matchen, Shannon Gray, Charles D. Fredrick, Tiffany Osburn, and Eric Shroeder. "Big Hole (41TV2161): Two Stratigraphically Isolate Middle Holocene Components in Travis County, Texas Volume I. *Index of Texas Archeology: Open Access Gray Literature from the Lone Star State* 2015, no. 1 (2016): 53.

Quigg, J. Michael, Paul M. Matchen, Charles D. Fredrick, and Robert A. Ricklis. "Eligibility Assessment of the Slippery Slope Site (41MS69) in TxDOT Right-of-Way in Mason County, Texas." *Index of Texas Archaeology: Open Access Gray Literature from the Lone Star State* 2015, no. 1 (2015): 9.

———. *Root-Be-Gone (41YN452): Data Recovery of Late Archaic Components in Young County, Texas, Volume I*. Austin: TRC Environmental Corporation, 2011.

Raff, Jennifer. *Origin: A Genetic History of the Americas*. New York: Hachette Book Group, 2022.

Raghavean, Maanasa, Matthias Steinrucken, Kelley Harris, Stephan Schiffels, Simon Rasmussen, Michael DeGiorgio, and Anders Albrechtsen. "Genomic Evidence for the Pleistocene and Recent Population History of Native Americans." *Science* 349, no. 6250 (2015): aab3884.

Raskevitz, Thornton, R. "Phytolith Analysis as a Paleoecological Proxy When Examining Bison Anatomical and Behavior Changes in the Great Plains." Master's thesis, Oklahoma State University, 2020.

Reinhard, Karl J., and Vaughn M. Bryant Jr. "Coprolite Analysis: A Biological Prospective on Archeology." *Papers in Natural Resources* (1992): 245–88.

Reinhard, Karl J., J. Richard Ambler, and Christine R. Szuter. "Hunter-Gatherer Use of Small Animal Food Resources: Coprolite Evidence." *International Journal of Osteoarchaeology* 17, no. 4 (2007): 416–28.

Reinhard, Karl, J. and Vaughn M. Bryant Jr. "Pathology and the Future of Coprolite Studies in Bioarchaeology." *Papers in Natural Resources, Paper 43* (2008): 210–15.

Reinhard, Karl J., Keith L. Johnson, Sara LeRoy-Toren, Kyle Wieseman, Isabel Teixeira-Santos, and Monica Viera. "Understanding the Pathological Relationship between Ancient Diet and Modern Diabetes through Coprolite Analysis: A Case Example from Antelope Cave, Mojave County, Arizona." *Papers in Natural Resources, Paper 321* (2012).

Reinhiller, Jayne. "Holding on to Culture: The Effects of the 1837 Smallpox Epidemic on Mandan and Hidatsa." *Butler Journal of Undergraduate Research* 4, no. 1 (2018): Article 12.

Resendez, Andres. *The Other Slavery: The Uncovered Story of Indian Enslavement in America*. Boston: Mariner Books, 2017.

Richard, Andrew J. "Clovis and Folsom Functionality Comparison." Master's thesis, University of Arkansas, 2015.

Richardson, Rupert N. "Neighbors, Robert Simpson." *Handbook of Texas Online*. www.tshaonline.org/handbook/entries/neighbors-robert-simpson.

Richardson, T. C. "Chisholm, Jesse." *Handbook of Texas Online*. www.tshaonline.org/handbookentries/chisholm-jesse.

Ricklis, R. A. "Archeology and Bioarcheology of the Buckeye Site (41VT98), Victoria County, Texas." *Coastal Environments, Inc. and U.S. Army Corps of Engineers, Galveston District* (2012): 28.

Ridley, Matt. *The Rational Optimist: How Prosperity Evolves*. New York: HarperCollins, 2010.

Rigby, Emma, Melissa Symonds, and Derek Ward-Thompson. "A Comet Impact in AD 536?" *Astronomy & Geophysics* 45, no. 1 (2004): 1–23.

Rivaya-Martinez, Joaquin. "A Different Look at Native American Depopulation: Comanche Raiding, Captive Taking, and Population Decline." *Ethnohistory* 61, no. 3 (2014): 391–418.
Robertson, R. G. "Rotting Face: Smallpox and the American Indian." *Journal of American History* 90, no. 1 (2003): 221–22.
Robinson, Jim. "Wildcat's Roar—Hooah!" *Orlando Sentinel*, April 1, 2001.
Roding, Christopher B. "Cherokee Towns and Calumet Ceremonialism in Eastern North America." *American Antiquity* 79, no. 4 (2014): 425–43.
Rodriquez, David Tovar. "A Possible Late Pleistocene Impact Crater in Central North America and Its Relation to the Younger Dryas Stadial." Master's thesis, University of Minnesota, 2020.
Roland, Paul. "Home on the Range: The Impact of the Cattle Trails on Indian Territory." *Armstrong Undergraduate Journal of History* 8, no. 2 (2018): 55–67.
Rothschild, Bruce M., Christine Rothschild, and Glen Doran. "Virgin Texas: Treponematosos-Associated Periosteal Reaction 6 Millenia in the Past." *Advances in Anthropology* 1, no. 2 (2011): 15–18.
Rosen, William. *Justinian's Flea: Plague, Empire, and the Birth of Europe*. New York: Viking Press, 2007.
Sabrin, Michael David. "Characterization of Acorn Meal." Master's thesis, University of Georgia, 2009.
Santiago, Adres Catalano, Juan Cesar Vilardi, Daniela Tosto, and Deatriz Ofelia Saidman. "Molecular Phylogeny and Diversification History of Prosopis (Fabaceae: Mimosoideae)." *Biological Journal of the Linnean Society* 93, no. 3 (2008): 621–40.
Sarris, Peter. "Climate and Disease" In *A Companion to the Global Early Middle Ages*, edited by Erik Hermans, 1–5. Yorkshire: Arc Humanities Press, 2020.
Schuetz, Arden Jean. *People-Events and Erath County Texas*. Stephenville, TX: Ennis Favors, 1970.
Seager, Richard, and Celine Herweijer. "Causes and Consequence of Nineteenth Century Droughts in North America." *Lamont-Doherty Earth Observatory of Columbia University: Drought Research*. ocp.Ideo.columbia.edu.
Seersholm, Frederik V., Daniel J. Werndly, Alicia Grealy, Taryn Johnson, Erin Keenan Early, Ernest L. Lundelius Jr., Barbara Winsborough, Grayal Earle Farr, Rickard Toomey, Anders J. Hansen, Beth Shapiro, Michel R. Waters, Gregory McDonald, Anna Lindeerholm, Thomas W. Stafford Jr., and Michael Bunce. "Rapid Range Shifts and Megafaunal Extinctions Associated with Late Pleistocene Climate Change." *Nature Communications* 11, no. 1 (2020): 2770.
Seiter, Thomas F. "Karankawas: Reexamining Texas Gulf Coast Cannibalism." Master's thesis, University of Houston, 2019.

Seymour, Deni J., "Comments on Genetic Data Relating to Athapaskan Migrations Implications of the Malhi Study for the Southwestern Apache and Navajo." *American Journal of Physical Anthropology* 139 (2009): 281–83. doi: 10.1002/ajpa.21062.

Shafer, Harry J., and Steve Tomka. "Early Middle Archaic Lithic Assemblages." In *An Early Middle Archaic Site along Cordova Creek in Comal County, Texas*, edited by Richard B. Mahoney, Harry J. Shafer, Steve A. Tomka, Lee C. Nordt, and Ramon P. Mauldin. *University of San Antonio Archaeological Survey Report*, 32 (2003): chapter 10.

Shankar, Uma. "Cave 'Krem Mawmluh' of Meghalaya Plateau—The Base of the 'Meghalayan Age' and '4.2 ka BP Event' in Holocene (Anthropocene)." *International Journal of Ecology and Environmental Sciences* 47, no. 1 (2021): 49–59.

Shuman, Bryan N. "Millennial Variations and a Mid-Holocene Step Change in Northern Mid-Latitude Moisture Gradients." *Research Square* (2021): 7–8.

Siepak, J., B. Walna, and S. Drzymala. "Speciation of Aluminum Released under the Effect of Acid Rain." *Polish Journal of Environmental Studies* 8 (1999): 55–58.

Silverton, Jonathan. *An Orchard Invisible: A Natural History of Seeds*. Chicago: University of Chicago Press, 2009.

Simon, Jean-Pierre. "Comparative Serology of a Disjunct Species Group: The Prosopis Julifora-Prosopis Chilensis Complex." *Aliso: A Journal of Systematic Evolutionary Botany* 9, no. 3 (1979): 483–97.

Simpson, Robert. *Neighbors and the Texas Frontier, 1836–1859*. Waco: Texian Press, 1975.

Sivad, E. Douglas. "Caballo, Jaun." *Handbook of Texas Online*. https://www.tshaonline.org/handbook/entries/caballo-juan.

Skinner, Christopher B., Juan M. Lora, Ashley E. Payne, and Christopher J. Poulsen. "Atmospheric River Changes Shaped Mid-Latitude Hydroclimate since the Mid-Holocene." *Earth and Planetary Science Letters* 541 (2020): 116–293.

Smith, F. Todd. *The Caddo Indians: Tribes at the Convergence of Empires, 1542–1854*. College Station: Texas A&M University Press, 1995.

———. *From Dominance to Disappearance: The Indians of Texas and the Near Southwest, 1786–1859*. Lincoln: University of Nebraska Press, 2005.

———. *The Wichita Indians: Traders of Texas and the Southern Plains, 1540–1845*. College Station: Texas A&M University Press, 2000.

———. "Wichita Locations and Population, 1719–1901." *Plains Anthropologist* 53, no. 208 (2008): 407–14.

Smith, Maria Ostendorf. "Treponemal Disease in the Middle Archaic to Early Woodland Periods of the Western Tennessee River Valley." *American Journal of Physical Anthropology* 131 (2006): 205–17.

Smithwick, Noah. *The Evolution of a State, Or, Recollection of Old Texas Days*, edited by Nanna Smithwick Donaldson. Austin: Gammel Book, 1900.

Sparks, Darrell. "Adaptability of Pecan as a Species." *HortScience* 40, no. 5 (2005): 1175–89.

Speth, John D., Khori Newlander, Andrew A. White, Ashley K. Lemke, and Lars E. Anderson. "Early Paleoindian Big-Game Hunting in North America: Provisioning or Politics?" *Quaternary International* 285 (2013): 111–39.

Stafford, Thomas W., and David L. Carlson. "The Age of Clovis—13,050–12,750 B.P." *Science Advances* 6, no. 43 (2020): eaaz0455.

Stahler, Daniel, and Brend Heinrich. "Common Ravens, *Corvus corax*, Preferentially Associate with Grey Wolves, *Canis lupis*, as a Foraging Strategy in Winter." *Animal Behavior* 64 (2002): 283–90.

Stambaugh, Michael C., Richard P. Guyette, Ralph Godfrey, E. R. McMurray, and J. M. Marschall. "Fire, Drought, and Human History Near the Western Terminus of the Cross Timbers, Wichita Mountains, Oklahoma, USA." *Fire Ecology* 5 (2009): 51–65.

Starzynska, Anna, Piotr Wychowanski, Maciej Nowak, Bartosz Kamil Sobocki, Barbara Alicja Jereczek-Fossa, and Monika Slupecka-Ziemilska. "Association between Maternal Periodontitis and Development of Systematic Diseases in Offspring." *International Journal of Molecular Sciences* 23, no. 5 (2022): 2473.

Stern, Peter. "Marginals and Acculturation in Frontier Society." In *New Views of Borderland History*, edited by Robert H. Jackson. Albuquerque: University of New Mexico Press, 1998.

Strong, Pauline Turner. "Transforming Outsiders: Captivity, Adoption, and Slavery Reconsidered." In *A Companion to American Indian History*, edited by Philip J. Deloria and Neal Salisbury, 339–56. Hoboken, NJ: Wiley-Blackwell, 2004.

Stucki, Devin S., Thomas J. Rodhouse, and Ron J. Reuter. "Effects of Traditional Harvest and Burning on Common Camas (*Camassia quash*) Abundance in Northern Idaho: The Potential for Tradition Resource Management in a Protected Area Wetland." *Ecology and Evolution* 11, no. 23 (2021): 16473–85.

Sun, Chijun, Timothy Shanahan, Pedro DiNezio, Nicholas McKay, and Priyadarsi Roy. "Great Plains Storm Intensity since the Last Glacial Controlled by Spring Surface Warming." *Nature Geoscience* 14 (2021): 912–17.

Sun, N., A. D. Brandon, S. L. Forman, M. R. Waters, and K. S. Befus. "Volcanic Origin for Younger Dryas Geochemical Anomalies ca. 12.900 cal BP." *Science Advances* 6, no. 31 (2020): eaax8587.

Sun, Nan, Alan D. Brandon, Steven L. Forman, and Michael Waters. "Geochemical Evidence for Volcanic Signatures in Sediments of the Younger Dryas Event." *Geochimica et Cosmochimica Acta* 321 (2021): 57–74.

Surovell, Todd A., Sarah A. Allaun, Barbara A. Crass, Joseph AM Gingerich, Kelly E. Graf, Charles E. Homes, and Robert L. Kelly. "Late Date of Human Arrival to North America: Continental Scale Differences in Stratigraphic Integrity of Pre-13,000 BP Archaeological Sites." *PloS One* 17, no. 4 (2022): e0264092.

Surovell, Todd A., Joshua R. Boyd, C. Vance Haynes Jr., and Gregory W. L. Hodgins. "On the Dating of the Folsom Complex and Its Correlation with the Younger Dryas, the End of the Clovis and Megafaunal Extinction." *PaleoAmerica* 2, no. 2 (2016): 81–89.

Swan, Susan, L. "Drought and Mexico's Struggle for Independence." *Environmental Review* 6, no. 1 (Spring 1982): 54–62.

Swanson, Donald A. "Coacoochee (Wild Cat)." *Handbook of Texas Online* (2024). www.tshaonline.org/handbook/entries/coacooochee-wild-cat.

Sweatman, Martin B. "The Younger Dryas Impact Hypothesis: Review of the Impact Evidence." *Earth Science Reviews* 218 (2021): 103677.

Tanner, Becky. "150 years Ago, Jesse Chisholm Opened Wichita's First Business." *Wichita Eagle*, January 27, 2013.

Taormina, Rebecca, Lee Nordt, and Mark Bateman. "Late Quaternary Alluvial History of the Brazos River in Central Texas." *Quaternary International* 631 (2022): 34–46.

Tarabek, Julianne E. "What's the Point: The Transition from Dart to Bow in the Eastern Plains." Master's thesis, University of Kansas, 2013.

Taylor, John F. "Sociocultural Effects of Epidemics on the Northern Plains." PhD dissertation, University of Montana, 1982.

Taylor, Matthew S. "The Midland Calvarium and the Early Human Habitation of the Americas." *Bulletin of the Texas Archeological Society* 86 (2015): 193–208.

Taylor, Stephen, Alvin Carl "A.C." Greene, Jr. *Handbook of Texas Online*. www.tshaonline.org/handbook/entries/greene-alvin-carl-jr-a-c.

Texas General Land Office. *Devoted to Peace: Delaware Chief John Conner*. Save Texas History, November 17, 2016.

Therrell, Matthew D., and Makayla J. Trotter. "Waniyetu Wowapi: Native American Records of Weather and Climate." *American Meteorological Society* (May 2011): 583–592.

Thompson, Jerry Don. "Colonel John Robert Baylor: Texas Indian Fighter and Confederate Soldier." *Hill Junior College Monograph*, no. 5 (1971).

Thoms, Alston V. "Ancient Savannah Roots of the Carbohydrate Revolution in South-Central North America." *Plains Anthropologist* 534, no. 205 (2008): 121–36.

———. "Ethnographies and Actualistic Cooking Experiments: Ethnaoarchaeological Pathways toward Understanding Earth-Oven Variability in Archaeological Records." *Ethnoarchaeology* 10, no. 2 (2018): 76–98.

———. "Rocks of Ages: Propagation of Hot-Rock Cookery in Western North America." *Journal of Archaeological Science* 36 (2009): 573–91.

Thoms, Alston V., ed.. "Archeological Survey at Fort Hood, Texas Fiscal Years 1991 and 1992: The Cantonment and Belton Lake Periphery Areas." *United States Army Fort Hood Archeological Resource Management Series, Research Report No. 27* (1993): 44.

Tippeconnic, Eric. "God Dogs and Education: Comanche Traditional Cultural Innovation and Three Generations of Tippeconnic Men." PhD dissertation, University of New Mexico, 2016.

Tomanek, G. W., and G. K. Hullet. "Effects of Historical Droughts on Grassland Vegetation in the Central Great Plains." In *Pleistocene and Recent Environments of the Central Great Plains*, edited by Wakefield Dort Jr., 205–9. Lawrence: University Press of Kansas, 1970.

Toohey, Matthew, Kirstin Kruger, Michael Sigl, Frode Stordal, and Henrik Svensen. "Climatic and Societal Impacts of Volcanic Double Event at the Dawn of the Middle Ages." *Climatic Change* 136 (2016): 401–12.

Troike, Rudolph C. "The Origins of Plains Mescalism." *American Anthropologist* 64, no. 5 (1962): 946–63.

Tucker, Phillip Thomas. "John Horse: Forgotten African-American Leader of the Second Seminole War." *Journal of Negro History* 77, no. 2 (1992): 74–83.

Tull, Delena. *Edible and Useful Plants of Texas and the Southwest*. Austin: University of Texas Press, 1987.

Turner, Ellen Sue, and Thomas R. Hester. *A Field Guide to Stone Artifacts of Texas Indians*. Houston: Gulf Publishing, 1985.

Turner, Matt Warnock. *Remarkable Plants of Texas: Uncommon Accounts of Our Common Natives*. Austin: University of Texas Press, 2009.

Turner-Pearson, Kathrine. "The Stone Site: A Waco Indian Village Frozen in Time." *Plains Anthropologist* 53, no. 208 (2008): 565–75.

Vagene, Ashild J., Alexander Herbig, Michael G. Campana, Nelly M. Robles Garcia, Christina Warinner, Susanna Sabin, and Maria A. Spyrou. "Salmonella Enterica Genomes from Victims of a Major Sixteenth-Century Epidemic in Mexico." *Nature Ecology & Evolution* 2, no. 3 (2018): 520–28.

Vanpool, Todd L. "The Survival of Archaic Technology in an Agricultural World: How the Atlatl and Dart Endured in the North American Southwest." *Journal of Southwestern Anthropology and History* 71, no. 3 (2006): 429–52.

Vierra, Bradley J., Nicholas Chapin, Christopher M. Stevenson, and M. Steven Shackley. "Another Look at Expedient Technologies, Sedentism, and the Bow and Arrow." *Kiva* 86, no. 4 (2020): 482–501.

Vines, Robert A. *Trees of Central Texas*. Austin: University of Texas Press, 1984.

Vitiello, Laura, Sara Ilari, Luigi Sansone, Manuel Belli, Mario Cristina, Frederica Marcolongo, and Carlo Tomino. "Preventive Measures against Pandemics from the Beginning of Civilization to Nowadays—How Everything Has Remained the Same Over the Millennia." *Journal of Clinical Medicine* 11, no. 7 (2022): 1960.

Vogel, Robert C. "Paul Laffite: A Borderland Life." *East Texas Historical Journal* 41, no. 1, Article 8 (2003).

Voosen, Paul. "Impact Crater under Greenland's Ice Is Surprisingly Ancient." *Science* 375, no. 65685 (2022): 1076–77.

"Waco Lake: Bosque River Basin and Beyond.'" *Texas Beyond History* (September 2009).

Wallace, Earnest, and E. Adamson Hoebel. *The Comanches: Lord of the South Plains*. Norman: University of Oklahoma Press, 1952.

Wanner, Heinz, Jurg Beer, Jonathan Butikofer, Thomas J. Crowley, Ulrich Cubasch, Jacqueline Fluckiger, and Hugues Goosse. "Mid- to Late Holocene Climate Change: An Overview." *Quaternary Science Reviews* 26, no. 19–20 (2008): 1791–828.

Wanner, Heinz, Olga Solomina, Martin Grosjean, Stefan P. Ritz, and Marketa Jetel. "Structure and Origin of Holocene Cold Events." *Quaternary Science Reviews* 30, no. 21–22 (2011): 3109–23.

Waters, Michael R. "Early Exploration and Settlement of North America During the Late Pleistocene." *The SAA Archaeological Record* 19, no. 3 (2019): 35–36.

———. "Late Quaternary Floodplain History of the Brazos River in East-Central Texas." *Quaternary Research* 43, no. 3 (1995): 311–19.

Waters, Michael R., Joshua L. Keene, Steven L. Forman, Elton R. Prewitt, David L. Carlson, and James E. Wiederhold. "Pre-Clovis Projectile Points at the Debera L. Friedkin Site, Texas—Implications for the Late Pleistocene Peopling of the Americas." *Science Advances* 4, no. 10 (2018): eaat4505.

Waters, Michael R. Joshua L. Keene, Elton R. Prewitt, Mark E. Everett, Tyler Laughlin, and Thomas W. Stafford Jr. "Late Quaternary Geology, Archaeology, and Geoarchaeology of Hall's Cave, Texas." *Quaternary Science Reviews* 274 (2021): 107276.

Watt, Frank H. "The Waco Indian Village and Its Peoples." *Texana* 6, no. 3 (Fall 1968): 195–237.

Weaver, J. E. "A Seventeen-Year Study of Plant Succession in Prairie." *American Journal of Botany* 41, no. 1 (January 1954): 31–38.

Webb, James. *Born Fighting: How the Scots-Irish Shaped America.* New York: Broadway Books, 2004.

Weddle, Robert S. *The San Saba Mission: Spanish Pivot in Texas.* Austin: University of Texas Press, 1964.

Wedel, Mildred M. "The Wichita Indians in the Arkansas River Basin." In *Plains Indian Studies: A Collection of Essays in Honor of John C. Ewers and Waldo R. Wedel,* edited by Douglas H. Ubelaker and Herman J. Viola, 118–34. Contribution to Anthropology, no. 30. Washington, DC: Smithsonian Institution Press, 1982.

Wedel, Waldo R. "Culture Sequence in the Central Great Plains." *Plains Anthropologist* 17, no. 57 (1972): 291–354.

Weist, Logan A., Don Esker, and Steven G. Driese. "The Waco Mammoth Monument May Represent a Diminished Watering-Hole Scenario Based on Preliminary Evidence of Post-Mortem Scavenging." *Palaios* 31, no. 12 (2016) 592–605.

Weniger, Del. *The Explorers Texas: The Animals They Found.* Vol. 2. Austin: Eakin Press, 1997.

Wernecke, D. Clark, and Michael B. Collins. "Patterns and Process: Some Thoughts on the Incised Stones from the Gault Site, Central Texas, United States." *Pleistocene Art of the World: Symposium* (2010): 670–78.

Whallon, R. E. "Social Networks and Information: Non-Utilitarian Mobility among Hunter-Gatherers." *Journal of Anthropological Archaeology* 25 (2006): 259–70.

Wilbarger, J. W. *Indian Depredations in Texas: Reliable Accounts.* Austin: Hutchings Printing House, 1889.

Wilkins, Barbara Throp. "Peter Garland: The Controversial Captain—Hero or Villain?" *Granbury Magazine*, Fall 1986.

Willerslew, Else, and David J. Meltzer. "Peopling of the Americas as Inferred from Ancient Genomics." *Nature* 594 (2021): 356–61.

Wilson, Diane Elizabeth. "The Paleoepidemiology of Treponematosis in Texas." PhD dissertation, University of Texas at Austin, 1988.

Wilson, Michael. "Geology in the Ice-Free Corridor." *A Journey to a New World.* Last updated 2005. Simon Fraser University. www.sfu.museum/journey/an-en/postsecondaire.

Wohlleben, Peter. *The Hidden Life of Trees: What They Feel, How They Communicate.* Vancouver: Greystone Books, 2015.

Wolfe, Kim. "Bur Oak (*Quercus macrocarpa* Michx.) in Riding Mountain National Park." Master's thesis, University of Manitoba, 2001.

Wong, Connie, and Jay L. Banner. "Holocene Climate Variability in Texas, USA: An Integration of Existing Paleoclimate Data and Modeling with a

New, High-Resolution Speleothem Record." *Quaternary Science Review* 127 (July 2015). doi:10.1016/j.quascirev.2015.06.023.

Wood, Gillen D'Arcy. "The Volcano That Changed the Course of History." *Conversation* (2014): 2–3.

Woodhouse, Connie A., and Jonathan T. Overpeck. 2000 Years of Drought Variability in the Central United States." *Bulletin of the American Meteorological Society* 79, no. 12 (December 1998): 2696.

Wrede, Jan. *Trees, Shrubs, and Vines of the Texas Hill Country*. College Station: Texas A&M University Press, 2015.

Yancy, William C. "*In Justice to Our Allies: The Government of Texas and Her Indian Allies, 1836–1867*." Master's thesis, University of North Texas, 2008.

Young, Dan M. *A Calendar of Erath County History*. Stephenville, TX: Vanderbilt Street Press, 1979–89.

———. "Identification of the Jumano Indians." Master's thesis, Sul Ross University at Alpine, 1970.

———. "Erath County." *Handbook of Texas Online*. www.tshaonline.org/handbook/entries/erath-county.

———. "Stephenville, TX." *Handbook of Texas Online*. www.tshaonline.org/handbook/entries/stephenville-tx.

Zhang, Chaun-rui, Amilia A. Dissanayake, and Muraleedharan G. Nair. "Functional Food Property of Honey Locust (*Gleditsia triacanthos*) Flowers." *Journal of Functional Foods* 18, Part A (October 2015): 266–74.

Zhang, Ming, Yonggang Liu, Jian Zhang, and Qin Wen. "AMOC and Climate Responses to Dust Reduction and Greening of the Sahara during the Mid-Holocene." *Journal of Climate* 34, no. 12 (2021): 4893–912.

Zesch, Scott. *The Captured: The True Story of Abduction by Indians on the Texas Frontier*. New York: St. Martin's Press, 2004.

Zimmerman, Kim Ann. "Cenozoic Era: Facts about Climate, Animals, and Plants." *Live Science* 9 (2016).

Zver, Lars, Borut Toskan, and Elena Buzan. "Phylogeny of Late Pleistocene and Holocene Bison Species in Europe and North America." *Quaternary International* 595 (2021): 30–38.

Index

Page numbers in *italic* refer to illustrations.

abortifacients, herbal, *102*, 103
acorns, 4, 7, 41–42, 47, 61; bur oak, 6; grinding, *27*; post oak, 3
Adobe Walls, Second Battle of. *See* Second Battle of Adobe Walls
adzes, prehistoric, 27
Afro-Native people, 117–20
agaves, 30–31, *31*, 34
agriculture, 63, 70, 71, 74, 77–79; Comanches, 122; Mexican Texas, 93; Wacos, 86
alcohol, 89, 115, 120, 123, 132
Altithermal period, 27–43, 73
American crow (*Corvus brachyrhynchos*), 6
American mastodon (*Mammut americanum*), 7, 11
American Revolution, 88
American robin (*Turdus migratorius*), 3
Anadarkos, 123–24, 125, 128, 180–81n54
Andice points, *37*, 37, 38
Anglo-Americans, 88–90, 92, 93, 105–6, 107, 113, 123; captives, 110, 112, 115; disguise, 137, 185n150; New Mexico, 136; smallpox, 109

Antelope Hills, Battle of. *See* Battle of Antelope Hills (Little Robe Creek)
Apaches, 40, 97, 136; trade, 96, 167n29. *See also* Lipan Apaches; Plains Apaches
Archaic period (North America), 23–56, 73
Arkansas yucca (*Yucca arkansana*), 30–31, *31*
arrowheads. *See* projectile points
Assiniboines, 93
atlatls, 15, 20, *21*, 27, 32, 36, 64
Austin, Moses, 89–90, 106–7
Austin, Stephen F., 90, 91–92, 107
Austin Phase (AD 700–1200), 69–70, 73

Bancroft, Hubert Howe, 126
barbed wire, 137–38
Bastrop, Felipe Enrique Neri, Baron de, 108
Battle, Nicholas, 126
Battle of Antelope Hills (Little Robe Creek), 124
Battle of Plum Creek, x, 112, 113
Battle of the Wichita Village, 124
Battle of Village Creek. *See* Village Creek Massacre
Battle of Walker's Creek, 113

Baylor, John R., 126–29
beads, bone. *See* bone beads
beans, 63, 78, 86, 93
beavers: dams, 44, 51, 69; prehistoric, 11, 139n5
Bell points, 37, *37*, 38
Beringia, 8–9
big bluestem (*Andopogon gerardi*), 27, 74
birds: feather use in arrows, 65, 67; as hunting guides, 100
bison, 35–37, 53–54, *55*, 134; Civil War drought and, 131; sinew use for bowstrings, 100; Toyah period, 74–75. *See also* giant bison (*Bison antiquus*)
bison hides and robes, 74, 85, 131
"black drink" (emetic), 82, 83
black-eyed pea (*Vigna unguicalata*), 86
black haw, southern. *See* rusty black haw (*Viburnum rufidulum*)
blackjack oak (*Quercus marliandica*), 6
Blair, John, 123
blue grama (*Bouteloua gracilis*), 28, 95
blue jay (*Cyanocitta cristata*), 6
bluestem grasses, 27, 28, 74, 95, *95*
bois d'arc (*Maclura pomifera*), 68–69, *68*, 74, 100, 161n43
bone beads, 20
bow and arrow, *58*, 63–65, 67–69, 74; Comanches, 95–96, *97*, 100
Brazos Indian Reservation, 117, 122, 125, 126, 127
Brazos River: arrow shafts, 65; Bosque confluence, 1, *10*, 86, 87; Coryell points, 70; Dark Ages, 59–60; Early Caddoans, 71; flooding and sediment stripping, 29, 52, 54; mustangs, 89; norteños, 108; Stephen Austin colony, 90, 91–92;

trade, 120; Wichitas, 77, 87. *See also* Brazos Indian Reservation
British troops, 114; Siege of Boston, 88
bubonic plague, 62
buffalo. *See* bison
buffalograss (*Bouteloua dactyloides*), 28, 30, 95
Buffalo Hump, 112, 116, 122, 124, 125
Burleson, Edward, 109
Burleson, Edward, Jr., 127
burned rock middens. *See* rock ovens
bur oak (*Quercus macrocarpa*), 6, 41
butchering, prehistoric, 15, 52, *53*
Buttermilk Creek, 12, 13, 18

C_3 grasses, 11, 19, 23, 27
C_4 grasses, 20, 27, 35, 36
Caballo Juan. *See* Horse, John (Gopher John)
Caddoans, 78, 79–80, 83, 111, 166n29; disease, 84; language, 77. *See also* Caddos; Early Caddoans; Tawakonis
Caddos, 88, 89, 108, 109, 113, 125; trade, 96, 167n29. *See also* Anadarkos
caffeine-based drinks, 82, 83
Cahokia and Cahokians, 71, 74, 78, 79
Calf Creek tradition, 35–39
calumet ritual, 80, 82, 83, 85
camas, 39, 51, 57, 58–59, 61, 69, 73; *Camassia sciloides*, 25, 48
camels, 6, 7, 11, 36, 68
Camp Cooper, 127, 182n84
Carmack, Joshua R., 126
Carolina moonseed. *See* snailseed (*Cocculus carolinus*)
Castroville points, 54, *54*, 57
cattail, southern (*Typha domingensis*). *See* southern cattail (*Typha domingensis*)

cattle, 117, 136, 137–38; Comanche raids, 132; Comanchero trade, 132, 133, 134. *See also* feral cattle
cedar waxwing (*Bobycilla cedrorum*), 3
ceramics. *See* pottery
Chaco Canyon and Chacoan people, 71, 78–79
Chatters, James C., 16
Cherokees, 89, 92–93, 111, 117
chert, vii, 16, 52, 72; knapping skills, 25–26, 36–38, 65; knapping tools, 48. *See also* projectile points
Cheyennes, 133, 135
Chisholm, Jesse, 117
Choctaws, 88, 89
Choctaw Tom, 125
cholera, 114, 122
cimarrones. *See* feral cattle
Civil War, 131, 132, 133
clams. *See* mussels
Clovis lithic tradition, 12–17
Coacoochee. *See* Wild Cat (Coacoochee)
cocoliztli epidemics, 84
Colt, Samuel, 113
Columbian mammoth (*Mammuthus columbi*), 7, 10
Comanche Creek Treaty. *See* Treaty of Comanche Creek
Comanche Indian Reservation, 122, 127
Comancheros, 132, 133, 134
Comanches, 47, 94–114, 129–37; Baylor and, 127; captives, 101–2, 104, 110, 111, 112; Council House Fight, x, 111–12, 116; French trade, 85; gift-giving, 108; horses and, 96, 98–99, 100, 103–5, 107, 109; Mexican Texas, 92, 93; Mexico, 122; Noconi, 110; Outbreak of 1874, 135; Quahadi, 131, 134, 136; raids, 101, 107–8, 110, 112, 122, 132; Red River War, 134, 135; smallpox, 96, 101, 109; Spanish Texas, 87; Sun Dance, 135, 184n135; treaties, 132; Wild Cat and, 119, 120–21; women, 102–4; Yamparika, 113. *See also* Northern Comanches; Penateka Comanches
common bean (*Phaseolus vulgaris*), 78
common persimmon (*Diospyros virginiana*), 47, *47*
common raven (*Corvus corax*), 100
Conner, John, 115, 116
contraceptives, herbal, *95*
controlled burns, 7
coprolites, 30, 32, 48, 77
corn, 62–63, 64, 78, 79, 80, 86
Cornett, Edward, 129, 131
Coronado, Francisco Vazquez de, 84
cotton, 138
cotton rats, 144n11
Council House Fight, x, 111–12, 116
cowpea (*Vigna unguiculate*), 86
coyote (*Canis latrans*), 14, 100
Coyote Droppings. *See* Isa-Tai (Coyote Droppings)
creation stories, 20, 81
Creeks (Muscogees), 111, 120, 179n22
Crees, 93
crisis cults, 134–35
Crockett, David, 119–20
crows, 6, 100

Dark Ages (AD 536–700), 59–63, 70
Darl points, 57, 59, 65, 160n38
dart points. *See* projectile points
datura (*Datura stramonium*), 80–81, *81*
deer, white-tailed. *See* white-tailed deer (*Odocoileus virginianus*)
deer bone fishhooks, 69–70, *69*
Delawares, 111, 112, 115, 121

dendrochronology, 3–4, 60
dental pathology of Early Peoples, 33, 34, 43, 56
depressions, 106; Great Depression, 138
DeShields, James T., 130–31
disease and diseases, 60, 84, 85, 96, 114, 122, 131; venereal, 126. *See also* pandemics; smallpox; treponematosis
dogwood, roughleaf. *See* roughleaf dogwood (*Cornus drummondii*)
drought and droughts, 52, 73, 94–95; Chacoan people and, 79; Civil War, 131, 132; "drought foods," 30–34; grasses and, 2; Great Depression, 138
drought-resistant plants, 28–30, 39
dyes, 79, 95

Eagle Drink (Quenah-evah), 132
Early Caddoans, 71–72
Early Triangular points, 26, *26*
Eastern Agricultural Complex (EAC), 77
Edwards, Hayden, 108
elephants, prehistoric, 142n31. *See also* mammoths
Ensor points, 57–58, *58*, 59
epidemics. *See* disease and diseases
Erath, George B., 123, 126
ethnic cleansing. *See* genocide
evolution, human. *See* human evolution
extinction of fauna, 19, 22, 35–36

farming. *See* agriculture
Feast of the First Fruits Ceremony, 83
feral cattle, 96–97
feral horses. *See* mustangs

fire, 2, 3, 6–7; grass fires, 7, 38, *40*, 74, 138, 174n49; honey mesquite and, 39, *40*. *See also* controlled burns
firearms, 85; Cherokees, 92; Red River War, 135. *See also* handguns; muskets; rifles
First Peoples, hunting by. *See* hunting by First Peoples
First Peoples (historical period), 74–93; agriculture, 63, 70, 71, 86; bow and arrow, *58*, 63–65, 67–69; calumet ritual, 80, *82*, 83, 85; camps, 63, 71, 75; crisis cults, 134–35; foods, 69–70, 77–78, 79, 86; genocide, 91, 105, 108, 109, 111; gift-giving, 108; government, 79; Little Ice Age, 74–83; Mexico, 84; slave trade, 101; smallpox, 60, 84, 85, 88, 93, 96, 101, 109, 114; tobacco use, 80; Toyah period, 74–76, 78–83; treaties, 92, 108–9, 115, 116, 117, 132; Woodland period, 63. *See also* Apaches; Caddoans; Cherokees; Cheyennes; Choctaws; Comanches; Creeks (Muscogees); Delawares; Karankawas; Kickapoos; Kiowas; Mississippian societies; Osages; Puebloans; reservations; Seminoles; Shoshones; Tonkawas; Waco people (Wacos); Wichitas
First Peoples (prehistory), 7–22, 81; camps, 51–52, 57, 59–60; creation myths, 20; dental pathology, 33, 34, 43, 56; foods, 15, 23–25, 29–34, 39–51; Middle Archaic, 27, 35–54, 56; stories and storytelling, 15–16. *See also* grinding tools; lithic technology
fish: in First Peoples' diet, 30, 48, 51, 69–70

Fisher, William, 111–12
fishhooks, deer bone, 69–70, *69*
fleas, 62
Florida, 118
flour, 31, 32, 41, *45*, *46*, 47, 66, 77
Folsom tradition, 20–22; Folsom points, *19*, 20–21
Ford, John "Rip," 124, 126, 127
Ford Alluvium, 73
Fort Parker. *See* Parker's Fort
fossils, 11; plants, 8, 143n4; pollen, 2, 11; wood, 1. *See also* coprolites
fox squirrel (*Sciurus niger*), 6
Fredonian Rebellion, 108
French traders and trade, 85–86, 87
fringed puccoon (*Lithospermum incisum*), 95, *103*
frostweed (*Verbesina virginica*), *102*, *103*
fungi, 4–7

Garland, Peter, 125, 126, 127
Gato del Monte. *See* Wild Cat (Coacoochee)
Gault site, 12–13, 16
genocide, 91, 105, 108, 109, 111
German colonists, 117, 136
giant bison (*Bison antiquus*), 7, 11, 12, *19*, 20, 23, *35*, *55*
giant cane. *See* river cane (*Arundinaria gigantea*)
Gilbreath, John Chesley, 137
Glacial Lake Agassiz. *See* Lake Agassiz
Godley points, 59
Gopher John. *See* Horse, John (Gopher John)
gopher tortoise (*Gopherus polyphemus*), 119
gourds, 78
Gower points, 26, 36

grain speculation, 106
grama, blue. *See* blue grama (*Bouteloua gracilis*)
grama, sideoats. *See* sideoats grama (*Bouteloua curtipendula*)
grama, Texas. *See* Texas grama (*Bouteloua rigidseta*)
Grant, Ulysses, S., 133, 134
grasses, 23, 27–28, 30, 77–78, 95, *103*, 138; Altithermal, 38; Austin Phase, 69, 73; Civil War drought, 131, 132; fires, 7, 38, *40*, 74, 138, 174n49; Little Ice Age, 95; maygrass, 78; use in baking, 25. *See also* C3 grasses; C4 grasses
"grass money" (term), 136
Great Depression, 138
greenbriers, 7
Grierson, B. H., 133
grinding tools, vii, 14, 17, 23, 27, 32, 40

hammerstones, vii, 75
handguns, 113–14, 123, 132
Harry, Jack, 118
Hays, John Coffee "Jack," 113, 116
herbal abortifacients. *See* abortifacients, herbal
herbal medicine. *See* medicine, herbal
herbal sterilization. *See* sterilization (birth control), herbal
herbs, psychoactive. *See* psychoactive plants
Hidalgo, Miguel, 106
high-fiber diet: Altithermal period, 33–34
hispid cotton rat (*Sigmodon hispidus*), 144n11
historiographic myth-making, 130–31
Homeric Grand Solar Minimum, 54, 156n26

honey locust (*Gledita triacanthos*), 45, 46–47
honey mesquite (*Prosopis glandulosa*), 6–7, 39–40, *40*, 42, 47
horned toad. *See* Texas horned lizard (*Phrynosoma cornutum*)
Horn Rock Shelter site, 14
Horse, John (Gopher John), 119, 120, 179n28
horses, 85; Apaches, 86; Comanches, 96, 98–99, 100, 103–5, 107, 109; fodder, 86; mounted fighting, 86, 96, 104–5, 114; mounted hunting, 100; prehistoric, 6, 7, 11, 15, 19, 22, 36; seed dispersal and germination, 6, *45*, *68*; Shoshones, 96; theft, 77, 87, 92, 98, 109, 118, 126–29 passim, 137; trade, 87, 96; value in mescal beans, 82
horses, feral. *See* mustangs
hot-rock baking, 23–25, 59, 70. *See also* rock ovens
Houston, Sam, 110, 113, 115, 116, 117, 120, 125, 129
Hoxie points, 26
human evolution, 16, 17, 25
hunting by First Peoples, 9, 11, 12–13, 15, 17, 19–22, 25, 65; Altithermal period, 32; bison, 20, 38, 74, 96, 99, 100; Calf Creek, 35–39; deer, 36, 70; feral cattle, 97; Shoshones, 94, 96

Indian disguise by non-Indigenous people, 127, 131, 137, 185n150
Indian Territory: Chisholm, 117; Jim Ned, 120; Jim Shaw, 116; Northern Comanches, 124; Seminoles, 119, 120; Texas Trail of Tears, 128–29
Indigenous peoples (historical period). *See* First Peoples (historical period)
Indigenous peoples (prehistory). *See* First Peoples (prehistory)
interstadials, 1, 2
Isa-Tai (Coyote Droppings), 134–35

jimsonweed. *See* datura (*Datura stramonium*)
John Horse. *See* Horse, John (Gopher John)
Johnson, Albert Sidney, 181n59, 182n84
Johnson, Middleton Tate, 129
José María, 123, 124, 127–28
Jumanos, 83, 85, 96, 166–67n29
Justinianic plague. *See* Plague of Justinian

Karankawas, 90, 91
Kelp Highway, 9, 11
Kichais, 77, 109
Kickapoos, 111, 118, 120, 131, 134
Kiowas, 113, 114, 122, 131, 133, 134, 136
knapping skills. *See* chert: knapping skills
knives, stone, 26, *53*, 72, 75

Lake Agassiz, 18
Lamar, Mirabeau, 110–13
lamb's-quarter (*Chenopodium berlandieri*), 77, 79
Last Glacial Maximum (LGM), 8–9
Lee, Robert E., 117, 181n59
Lehmann, Herman, 123, 184n143
Lenape. *See* Delawares
León, Alonso de, 96
Leon River, 70, 71, 142n31
limestone, vii, 1, 15
Lipan Apaches, 86–87, 96, 112, 122–23, 134
liquor. *See* alcohol

lithic technology, 12–22, 36; knapping tools, 48; Toyah period, 75. *See also* hammerstones; projectile points
Little Arkansas, Treaty of. *See* Treaty of Little Arkansas
little barley (*Hordeum pusillum*), 77–78
little bluestem (*Schizachyrium scoparium*), 28, 95, 95
Little Ice Age (AD 1300–1850), 74–76, 85, 95, 105
Little Robe Creek, Battle of. *See* Battle of Antelope Hills (Little Robe Creek)
live oak (*Quercus virginiana*), 22, 41
Lockhart, Matilda, 111
Louisiana Territory, 87, 88–89, 105–6
low-calorie diet: Altithermal period, 33–34

Mackenzie, Ranald S., 134, 135, 136
mammals, prehistoric, 7, 8–9, 11, 12, 19, *45*
mammoths, 7, *10*, 11, 12, 19
Mandans, 93
manos and metates, vii, 23, 25, *27*, 52, 76
María, José. *See* José María
Marlin, William, 128
maroons, 179n22
mastodon, American. *See* American mastodon (*Mammut americanum*)
matriarchal societies, 17, 99, 102
Maunder Minimum, 85
McLennan, Neil, 123
McNeill, W. W., 125–26
meat: in First Peoples' diet, 15, 25, 48, 49; trade, 74 *See also* butchering, prehistoric
Medieval Warm Period, 70–72, 73, 78, 94

medicine, herbal, 67, 95; prehistoric, 30, 39, 47. *See also* abortifacients, herbal; sterilization (birth control), herbal
Medicine Lodge, Treaty of. *See* Treaty of Medicine Lodge
megafauna, 7, 11, 15, 142n31; extinction, 19, 22, 36. *See also* mammoths
"Meghalayan Stage" (term), 52
mescal bean. *See* Texas mountain laurel (*Sophora secundiflora*)
mesquite trees, 68, 105. *See also* honey mesquite (*Prosopis glandulosa*)
metates. *See* manos and metates
Meusebach, John O., 117
Mexico: Comanche raids, 107–8, 122; Fourth Cavalry (US), 134; John Conner, 115, 116; Mexican-American War, 122; Mexican Texas, 92–93, 106–7, 108; Mixtecs, 84; slave market, 101; Stephen Austin, 90, 92; War of Independence, 106; Wild Cat, 120. *See also* Texas Revolution
Mézières, Athanase de, 87, 88
Mier y Terán, Manuel de, 108–9
Mississippian societies, 64, 72, 79, 80, 83. *See also* Cahokia and Cahokians
Mixtecs, 84
mollusks, 48–51
Moore, John H., 109, 112, 117
Mount Tambora eruption (1815), 105, 106, 114
Muguara, 111–12
Mule Creek Massacre. *See* Pease River Massacre
mules, 107
Murphy, Patrick, 127, 129, 131
Muscogees. *See* Creeks (Muscogees)

muskets, 86, 87, 91, 104, 167n4
mussels, 50–51
mustangs, 89, 96–98, 99, 107
mycelia, 4–7
myth-making, historiographic. *See* historiographic myth-making

Native Americans. *See* First Peoples (historical period)
Ned, Jim, 117–18, 120
Neighbors, Robert Simpson, 121–22, 124, 126, 127, 128, 129
Neil, James, 93, 109
Nelson, Allison, 126
New Mexico, 96, 107, 132, 136; Pueblo Revolt of 1680, 85, 96
Nolan, Philip, 89, 106
Nolan points, 26–27, 32, 42, *42*
Northern Comanches, 124
nut and seed dispersal. *See* seed and nut dispersal

oaks, 6, 22, 41–42; masting, 4, 41. *See also* acorns; post oak (*Quercus stellata*)
ocher, red. *See* red ocher
opuntia, 24, 29–30, *29*, 34, 72
Osage orange. *See* bois d'arc (*Maclura pomifera*)
Osages, 77, 86
ovens, rock. *See* rock ovens

paleofeces. *See* coprolites
Paleo-Indians. *See* First Peoples (prehistory)
Palmer, T. N., 126
pandemics, 61–62, 64, 93, 114; tree diseases, 5
Panic of 1819, 89, 106
parasites, 32, 34

Parker, Cynthia Ann, 101, 110, 130
Parker, John (1758–1836), 109
Parker, John Richard, 110
Parker, Quanah, 130, 135, 136
Parker, Silas M., 109
Parker, William B., 116
Parker's Fort, 109
peach trees, 86, 122
Pease River Massacre, 130–31
pecan (*Carya illinoinensis*), 6, 28, 44–46
Pedernales points, 53, *53*, 54
Penateka Comanches, 106, 108, 111–12, 113, 115, 122; Jim Ned and, 118; Texas Trail of Tears, 128–29
Perdiz points, 75, *76*
persimmon, common (*Diospyros virginiana*). *See* common persimmon (*Diospyros virginiana*)
peyote (*Lophophora williamsii*), 85
phytoliths, 11, 144n11
pipe ritual. *See* calumet ritual
pistols, 113–14; Comanchero trade, 132
Plague of Justinian, 61–62
Plains Apaches, 77, 78, 83, 96
Plains Crees, 93
Plains Village tradition, 74, 78
plants, medicinal. *See* medicine, herbal
plants, psychoactive. *See* psychoactive plants
Pleistocene, 1–2, 6, 7, 12, 30; adzes, 27; Last Glacial Maximum (LGM), 8–9; Younger Dryas event, 17–20, 22
Plum Creek, Battle of. *See* Battle of Plum Creek
Plummer, J. B., 127
Poinsett, Joel, 92, 170n42
post oak (*Quercus stellata*), 2–4, 5–6, 23, 28, 41, 71

pottery, 40, 72, 74, 75; Intermountain Ware, 96
prescribed burns. *See* controlled burns
prickly pears: opuntia subspecies. *See* opuntia
projectile points, 25–27, 36–38, 52–53, 54, 57–59, 65; Bell and Andice, 37–38, *37*; Bonham-Alba, 72; Castroville, 54, *54*, 57; Clovis, 12–16, *14*; Coryell, 70; Darl, 57, 59, 65, 160n38; Early Triangular, 26, *26*; Ensor, 57–58, *58*, 59; Folsom, *19*, 20–21; Godley, 59; Gower, 26, 36; Harrell, 75; Hoxie, 26; Marcos, 54, *54*; Marshall, 54, 57; Nolan, 26–27, 32, 42, *42*; Pedernales, 53, *53*, 54; Perdiz, 75, *76*; Scallorn, *58*, 64, 69, *69*, 70, 71, *71*, 73, 74; Shoshone/Comanche, 95–96, *97*; Uvalde, 26, 36; Washita, 74, 75, *75*
pronghorn (*Antilocapra americana*), 48, *49*
psychoactive plants, 80–83, 85
puccoon, fringed (*Lithospermum incisum*). *See* fringed puccoon (*Lithospermum incisum*)
Puebloan peoples, 84, 85; revolt of 1680, 85, 96; trade, 96, 166n29

Quahadi Comanches, 131, 134, 136
Quenah-evah. *See* Eagle Drink (Quenah-evah)

Rabdotus mooreanus, 49–50
rape, 101, 108, 109, 110, 129
rats, 32, 62, 144n11
ravens, 100
red haw (*Crataegus phaenopyrum*), 98 Red Jack, 123, 180–81n54
red ocher 13, 14, 146n18

Red River, 85, 86, 87; bois d'arc, *68*; Jim Shaw, 116; Plains Villagers, 74; Texas Trail of Tears, 128–29; trade, 85, 86, 107; Wild Cat, 119
Red River War, 134, 135
Republic of Texas, 110–13; Chisholm, 117; Council House Fight, x, 111–12, 116
reservations, 132. *See also* Texas Indian reservations
rifles, 91, 92–93, 112, 119, 135
river cane (*Arundinaria gigantea*), 65, *66*
robin, American (*Turdus migratorius*). *See* American robin (*Turdus migratorius*)
rock ovens, 17, 23–25, *24*, 39, 40, 48, 70
Roman Solar Minimum, 61, 63
Ross, Lawrence Sullivan "Sul," 129, 130
Ross, Ruben, 105
Ross, Shapley P., 125
roughleaf dogwood (*Cornus drummondii*), 65, *67*, 100
Runnels, Hardin, 124, 126, 128
rusty black haw (*Viburnum rufidulum*), 97–98

San Antonio, 87, 111, 112, 126, 129; compulsory immunization, 109; Council House Fight, x, 111–12, 116; Moses Austin, 89, 107
sandstone, 1; metates and manos, vii, 23, 25, 27
San Gabriel River, 109, 174n47
Scallorn points, *58*, 64, 69, *69*, 70, 71, *71*, 73, 74
Scots Irish, 89, 90–92, 93, 94, 107
Second Battle of Adobe Walls, 135
"second harvest" (seed reuse), 32
seed and nut dispersal, 6, *45*, *68*

seeds in human diet, 31, 32, 33, 40, 46, 66, 77; beverage use, 82–83
Seminoles, 111, 118–19, 120, 179n22
Shaw, Bill (Tall Man), 115, 116, 117
Shaw, Jim (Bear Head), 115, 116–17
Shawnees, 111, 115, 121
Sheridan, Philip, 133
Sherman, William Tecumseh, 118, 133–34
Shoshones, 94, 99, 101; arrowpoints, 95–96, 97; Battle of Walker's Creek, 113; horses, 96; smallpox, 96
sideoats grama (*Bouteloua curtipendula*), 28, *28*, 95
slavery and enslaved people, 101, 117, 123; Book of Mormon story, 173n37; slave raiding, 96, 117
smallpox, 60, 84, 85, 88, 93, 109, 114; Comanches, 93, 96, 101, 109; Shoshones, 96
snails, 49–50
snailseed (*Cocculus carolinus*), 7
soaps and shampoos, prehistoric, 31, *31*
solar minima, 54, 60, 61, 63, 74, 85, 156n26
southern cattail (*Typha domingensis*), 46, 47–48, 57
southern live oak. *See* live oak (*Quercus virginiana*)
southern wild rice (*Zizania aquatica*), 77, 78
Spanish colonization, 87–89, 96–97, 106; slave market, 101
spiny greenbrier (*Smilax bona-nox*), 7
Spiro culture, *68*, 71, 74, 79
squash, 63, 78, 86
squirrels, 4, 6, 32; hide use, 69
spruce trees, 2
Stephen, James, 123
Stephen, Samuel, 125

Stephenville, Texas, 123, 137
sterilization (birth control), herbal, 95, 103
stone tools. *See* lithic technology
sumpweed (*Iva annua*), 77
Sun Dance, 114, 135, 184n135

Tahbynaneekah (Hears the Sunrise), 133
Taovayas, 87, 92
Tarrant, Edward, 113
tattooing, 14
Tatum, Lawrie, 133
Tawakonis, 76, 86, 87, 92–93, 109
Texas grama (*Bouteloua rigidseta*), 95, *103*
Texas horned lizard (*Phrynosoma cornutum*), 32, 100
Texas Indian reservations, 121–23; Brazos Reserve, 117, 122, 125, 126, 127; Comanche, 122, 127
Texas mountain laurel (*Sophora secundiflora*), 82–83
Texas prickly pear (*Opuntia engelmannii*, var. *lindheimeri*), 29–30, *29*
Texas Rangers, 90, 93, 113, 117, 121; John Ford, 124, 126, 127; Pease River Massacre, 130–31
Texas Revolution, 110, 115, 116
Texas thistle (*Cirsium texanum*), 31, *33*
Texas Trail of Tears, 128–29
tobacco, 80
Tonkawas, 112, 123
tortoises, 119
Toyah period (1250–1700), 74–76, 78–83
trade: alcohol, 89; Anadarkos, 123; Comancheros, 132, 133, 134; French, 85–86, 87; Plains Village, 74; post-Archaic, 64, 68, 70, 72, 74,

78, 85–89, 96, 120; prehistoric, 12, 13–14, 21
Trail of Tears (Texas). *See* Texas Trail of Tears
treaties, 92, 108–9, 115, 116, 117, 132
Treaty of Comanche Creek, 117
Treaty of Little Arkansas, 132
Treaty of Medicine Lodge, 132
trees, 2, 6; Altithermal, 28, 38; Ashe juniper ("cedar"), 138; Chisholm, 117; cottonwood, 104; Early Holocene, 22; Late Archaic, 44; mycelia and, 5–7; recruitment pulse, 4; spruce, 2. *See also* bois d'arc (*Maclura pomifera*); dendrochronology; mesquite trees; oaks; peach trees; pecan (*Carya illinoinensis*)
treponematosis, 62, 63, 64
Trinity River, 89, 93, 97, 108, 109

US Army, 133–34; Camp Cooper, 127, 182n84; First Infantry, 127; Florida, 119, 120; Fourth Cavalry, 134; Red River War, 134; Second Cavalry, 124, 126, 129, 130, 181n59; Seminoles and, 119, 120
US-Mexico relations, 92, 122
Utes, 101
Uvalde points, 26, 36

Van Dorn, Earl, 124, 125, 129
vigilantes and vigilantism, 125–28, 137
Village Creek Massacre, 113
volcanic eruptions, 18, 60–61, 63; Little Ice Age, 74, 105–6, 114

Waco people (Wacos), 76, 86, 87, 92–93, 109, 118
Walker, Samuel H., 113

Walker's Creek, Battle of. *See* Battle of Walker's Creek
War of 1812, 89, 105
Washington hawthorne. *See* red haw (*Crataegus phaenopyrum*)
Washita points, 74, 75, *75*
Washita River, 129
Western, Thomas G., 120
West Range Alluvium, 73
whiskey, 89, 115, 120, 123; Comanchero trade, 132
white-tailed deer (*Odocoileus virginianus*), 48, 72; antler use for tools, 65; bone fishhooks, 69–70, 69; sinew use for bowstrings, 100
Wichitas, 76–89, 91–92, 109, 113, 123; Chisholm and, 117; Jim Ned and, 118; Peace of 1827, 108; trade, 96, 167n29. *See also* Taovayas
Wichita Village, Battle of the. *See* Battle of the Wichita Village
Wild Cat (Coacoochee), 118–19, 120, 179n28
wild rice, 77, 78
wild tobacco. *See* tobacco
Wolff's Law, 17
Wolf Solar Minimum, 64, 74
Woodland period, 63
woodworking tools, prehistoric, 27
wolves, 100
woolly mammoth (*Mammuthus primigenius*), 10

yaupon (*Ilex vomitora*) and yaupon tea, *82*, 83
Younger Dryas period, 17–20, 22, 35–36
yucca, 30–31, *31*, 69